Assessing Compensation Reform

Research in Support of the
10th Quadrennial Review
of Military Compensation

Beth J. Asch, James Hosek, Michael Mattock, Christina Panis

Prepared for the Office of the Secretary of Defense
Approved for public release; distribution unlimited

NATIONAL DEFENSE RESEARCH INSTITUTE

The research described in this report was prepared for the Office of the Secretary of Defense (OSD). The research was conducted in the RAND National Defense Research Institute, a federally funded research and development center sponsored by the OSD, the Joint Staff, the Unified Combatant Commands, the Department of the Navy, the Marine Corps, the defense agencies, and the defense Intelligence Community under Contract W74V8H-06-C-0002.

Library of Congress Cataloging-in-Publication Data is available for this publication.

ISBN 978-0-8330-4567-6

The RAND Corporation is a nonprofit research organization providing objective analysis and effective solutions that address the challenges facing the public and private sectors around the world. RAND's publications do not necessarily reflect the opinions of its research clients and sponsors.

RAND® is a registered trademark.

Published 2008 by the RAND Corporation
1776 Main Street, P.O. Box 2138, Santa Monica, CA 90407-2138
1200 South Hayes Street, Arlington, VA 22202-5050
4570 Fifth Avenue, Suite 600, Pittsburgh, PA 15213-2665
RAND URL: http://www.rand.org/
To order RAND documents or to obtain additional information, contact
Distribution Services: Telephone: (310) 451-7002;
Fax: (310) 451-6915; Email: order@rand.org

Preface

Military compensation is a pillar of the all-volunteer force. It is a fundamental policy tool for attracting and retaining personnel, and its structure—and the incentives implied by that structure—can affect U.S. service members' willingness to join, exert effort, demonstrate their leadership potential, remain in the military, and, eventually, exit the military at an appropriate time. Military compensation is a composite of current pay and allowances, special and incentive pays, health benefits, disability benefits, retirement benefits, and other benefits. Its importance to the readiness and morale of the force is such that it is reviewed every four years to determine whether its form and amounts are adequate to meet manpower objectives.

This monograph presents research undertaken for the 10th Quadrennial Review of Military Compensation (QRMC), and the focus of the research is on the mix and structure of current versus deferred compensation. RAND was asked to analyze several policy alternatives under consideration by the 10th QRMC that would change the military retirement-benefit system and add elements to current compensation. This monograph summarizes the results of the analysis of those options of interest to the 10th QRMC.

This monograph should be of interest to audiences concerned with the sustainability of the all-volunteer force, the relationship between compensation structure and the experience and grade structure of the personnel force, the potential costliness and cost savings of the policy alternatives relative to current compensation, and the modeling of complex job-tenure decisions in circumstances in which current choices affect future opportunities and the future may be uncertain. It considers these questions in a context that integrates active-duty retention decisions with post-active-duty decisions to participate in the selected reserve.

This research was conducted within the Forces and Resources Policy Center of the RAND National Defense Research Institute, a federally funded research and development center sponsored by the Office of the Secretary of Defense, the Joint Staff, the Unified Combatant Commands, the Department of the Navy, the Marine Corps, the defense agencies, and the defense Intelligence Community.

For more information on RAND's Forces and Resources Policy Center, contact the Director, James Hosek. He can be reached by email at James_Hosek@rand.org; by phone at 310-393-0411, extension 7183; or by mail at the RAND Corporation, 1776 Main Street, Santa Monica, California 90407-2138. More information about RAND is available at www.rand.org.

Contents

Figures

Tables

Summary

Military retirement reform has been a central element of the policy debate regarding why and how to restructure the system for compensating members of the U.S. armed forces. Concerns about the compensation system, and the retirement system specifically, include the rising cost of military compensation and the need for greater efficiency in the provision of compensation, the greater need for flexibility to reshape the force as missions change in ways that challenge the current compensation system, and issues related to the equity of military retirement benefits of active versus reserve personnel, junior versus senior personnel, and military personnel versus their civilian counterparts. Active members can claim retirement benefits before reservists can; junior members who leave prior to completing 20 years of service do not qualify for retirement benefits, unlike their more senior counterparts; and the 20-year vesting rule is outside the civilian vesting norm of five to seven years of service, under the Employee Retirement Income Security Act of 1974 (ERISA) (P.L. 93-406).

The 10th Quadrennial Review of Military Compensation (QRMC), building on previous studies and commission reports, including the 2006 report of the Defense Advisory Committee on Military Compensation (DoD, 2006) and the 2000 report of the Defense Science Board Task Force on Human Resources Strategy, has proposed an alternative military retirement system that addresses concerns regarding the current system while still sustaining the force. The new system would include the following:

- A defined benefit (DB) plan with earlier vesting at year of service (YOS) 10: The DB plan would provide an annuity at age 57 equal to 2.5 percent times YOS times high-three annual basic pay (ABP).
- An early-withdrawal option for the DB plan: Those with 20 or more years of service could choose to receive an immediate annuity, rather than waiting until age 57, but the annuity would be reduced by 5 percentage points for every year that the service member is less than 57 years old. That is, the reduction factor is $0.05 \times (57 - age)$.
- A defined contribution (DC) plan, vested at YOS 10: Under this plan, the U.S. Department of Defense (DoD) would contribute to a fund for each member, up to 5 percent of basic pay, depending on YOS; members would own the fund once they are vested. In this sense, the benefit is portable. The payout of the DC plan is according to 401(k) rules. Under these rules, the payout begins at age 59.5, though in our analysis, we assume that it is age 60.
- "Gate pay": This is a multiple of basic pay and would be paid to those who complete specific milestones, i.e., a specific number of years of service, regardless of whether they stay or leave upon completion. The specific number of years of service will depend on the retention pattern and force shaping that the service would like to achieve, as well as the current force shape.
- Separation pay, vested at YOS 20, would be provided to members who separate, after they leave the service. The formula is a multiple of monthly basic pay (MBP) times YOS.

The DB and DC plans are the foundation of the alternative system that we consider in this analysis. Gate pay and the level of separation pay depend on the force-shaping objectives of the service, and they can vary across the services or even within a service among personnel in different communities, such as occupational groups. For example, gate pay and separation pay can be set to achieve longer careers among those in more technical occupations or among specific officer communities.

RAND was asked to develop a modeling capability to assess compensation alternatives, such as the QRMC proposal, in terms of their effects on military retention, retirement behavior, vesting, cost, reserve participation, and the value of compensation from the perspective of the member leaving active duty. This monograph presents the results of that study. It reviews the case of military compensation reform, documents the model, and provides an assessment of the QRMC proposal.

To assess the proposal, we focused on several cases. The first case includes the DC and DB plan and sets gate and separation pay to enable each service to sustain its personnel strength and achieve the same retention profile as produced by the current compensation system. The second case also includes the DC and DB plan and sets gate and separation pay to enable the services to achieve longer careers, e.g., to enable more members to stay beyond YOS 20. The specific profiles that we examined were based on guidance from the QRMC director. To illustrate the range of variability in career profiles that might be produced by the QRMC alternative for different communities, our other cases vary gate and separation pay to generate a shorter career than the current profile and a substantially longer career than the long career recommended by the QRMC.

Model Development

The model that we developed builds on the Gotz-McCall dynamic retention model for active-duty personnel and extends it to the reserves (see Gotz and McCall, 1984). The model is a stochastic dynamic programming model of the decision to stay or leave active duty and, if a member leaves, the decision to participate or not participate in the reserve components in the subsequent periods. The dynamic retention model is formulated in terms of the parameters that underlie the retention and reserve participation processes. Because the model is based on the factors underlying the decision process, rather than on a specific compensation system and retention outcome, the model permits assessments of alternative compensation systems. We estimate seven model parameters—namely, the means, variances, and covariance of the preference for active and reserve service (which account for five parameters), a parameter related to the variance of the stochastic shock affecting the decision to stay on active duty or leave, and a parameter related to the variance of the stochastic shocks affecting the alternatives of being a civilian or a reservist. We estimate the model for each service for the enlisted force using the Work Experience File (WEX) from 1990 to 2007, provided by the Defense Manpower Data Center. The WEX data track the careers of service members in the active and reserve components; for our analysis, we focused on enlisted members. We supplement the WEX data with information on military compensation, as well as data on civilian earnings from the Current Population Survey (U.S. Census Bureau and U.S. Department of Labor Statistics, various years).

Given the model parameter estimates, we simulate retention profiles, as well as other outcomes of interest, such as retirement vesting, cost, and reserve participation under the current compensation system and the QRMC alternative. We found that we could closely replicate existing active-duty retention patterns, including the percentage of service members who reach active-duty retirement eligibility at YOS 20.

Results

To replicate the current force under the QRMC retirement alternative, gate pay, equal to 15 percent of ABP, is offered at YOS 12 and 18, and separation pay, equal to MBP times YOS, is offered to those with between 20 and 24 years of service (in the Army). To extend careers beyond YOS 20 along the lines suggested by the QRMC, gate pay is set equal to 25 percent at YOS 12 and 35 percent at YOS 18, and separation pay is MBP times YOS for those with between 20 and 30 years of service (in the Army).

We found that the QRMC alternative can reproduce the retention patterns achieved under the current system by the appropriate setting of gate and separation pay, on top of the DB and DC plans. The QRMC alternative achieves the same rate of retirement for an entering cohort, the same midcareer retirement patterns, and the same manyears per accession among the active force. To offer a comparison, under the current system, our data show that 10.5 percent of Army entrants reach YOS 20 and that the average man-years per accession is 7.0. Under the QRMC proposal, our estimates show that 10.8 percent reach YOS 20, and average man-years per accession is 7.1. We achieved similar results for the other services.

The QRMC alternative can also reshape the force by inducing longer careers—specifically, higher retention at YOS 20—while sustaining force levels, as suggested by the QRMC. For example, by appropriately setting gate and separation pay, the QRMC retirement alternative can increase the Army enlisted retirement rate to 12.6 percent, and average man-years per accession increase to 7.6 years. Again, similar results are found for the other services, though the levels of gate and separation pay vary by service.

The QRMC proposal also provides the flexibility to conduct force shaping within a service (i.e., gate and separation pay can be varied within a service to produce different retention profiles for different communities in that service, such as different occupational groups). We illustrate the potential to vary profiles within the Army by eliminating gate pay and vesting personnel at YOS 10 for separation pay. This profile produces a shorter career at less cost than under the current system. We also produced a profile with greater retention beginning in the early career and continuing through the end of the career by vesting separation pay at YOS 20 and offering gate pay equal to 40 percent at YOS 12, 14, 16, and 18. The substantially longer career increases retention. We find that, supposing that the Army wanted to retain one-third of personnel who fit the current career profile, one-third of personnel with a short career, and one-third of personnel with a significantly longer career, the weighted cost per active man-year is slightly lower than that under the current system.

The QRMC proposal is less costly than the current system, given the gate and separation pays we considered. For the Army, the QRMC achieves current retention patterns and force structure at 6.1-percent lower cost in terms of active-duty cost per man-year, where cost includes the current cost of regular military compensation, gate pay, and separation pay, plus the outlays required to fund the DB and DC plans for vested personnel when they leave service. Also, the QRMC alternative achieves the longer career profile that we considered for the

Army force at 3-percent lower cost. Thus, the system is more efficient than the current system in achieving a given force structure. The reason for the improvement in cost-effectiveness is that the QRMC system shifts compensation away from the end of the career, in the form of retired pay, and toward the earlier part of the career, in the form of gate and separation pay. Furthermore, these pays are targeted to those who reach specific career milestones, unlike a basic-pay increase that would be received by all personnel.

The QRMC proposal vests personnel earlier. The DB part of the QRMC proposal vests at YOS 10, versus YOS 20 under the current system, and the DC part of the QRMC proposal vests at YOS 10. We find that the percentage of entrants who vest more than doubles for the Army. In the first case, in which the QRMC proposal replicates the current Army enlisted force, 23.7 percent vest at YOS 10, which compares with 10.5 percent who vest at YOS 20 under the current system. In the second case, in which the QRMC proposal extends active-duty careers, the percent vesting at YOS 10 versus YOS 20 increases from 10.5 percent to 25.2 percent. Thus, more individuals become eligible for retirement benefits under the QRMC proposal.

Because the QRMC system significantly restructures the retirement system, the amount and timing of retirement benefits change. In the case in which the same force profile is achieved, assuming that individuals choose the early-withdrawal option, we find that the QRMC alternative increases compensation for leaving members, given our assumptions about the personal discount rate. For example, actuarial tabulations show that an E-7 who leaves at YOS 20 would receive a present discounted value of $120,000 under the current retirement-benefit system, assuming a personal discount rate of 15 percent. Under the QRMC alternative, an E-7 who leaves at YOS 20 and takes the early-withdrawal option for the DB plan would receive $138,000, including the values of their DB and DC plans at that point, plus separation pay and gate pay. A similar result was found for members leaving at YOS 10, YOS 24, and YOS 30. On the other hand, if the member opts to defer the DB annuity until age 57, i.e., if the member does not take the early-withdrawal option, the QRMC alternative provides less compensation than the current system in some cases. Clearly, for a member leaving at YOS 10, the QRMC plan provides more money because the current system provides no benefit for such a member. We also find that the QRMC plan provides a greater benefit to those leaving at YOS 20, but we do not find this to be the case for those leaving later, e.g., at YOS 30.

Our findings are tempered by the fact that they depend on our assumptions. Specifically, we assume that gate pay and separation pay will be the same for those taking the deferred option versus the early-withdrawal option. If gate pay or separation pay is increased in the former case, we could find that compensation is higher for those who leave later than YOS 20, even under the deferred option. Of course, cost would increase too. More generally, the value to the individual of the current system versus the QRMC proposal depends on the individual's personal discount rate. This rate no doubt varies among individuals. Individuals with a higher personal discount rate are more likely to favor the QRMC alternative, under which compensation is more front-loaded, whereas the reverse is true for individuals with a lower personal discount rate. According to our actuarial tabulations, at a personal discount rate below 12.5 percent, the QRMC alternative is generally less attractive than the current system.

On the basis of our findings, we conclude that the QRMC alternative has the potential to address the key concerns about the current retirement system. Our analysis suggests that it would be more cost-effective, increase the equity of the system, and enable force-management initiatives to reshape the force to suit changing requirements or alternative needs of personnel throughout the force or in specific communities. Changing to a new compensation system is not easy, and additional questions remain about the advisability of such a change. The analysis presented here contributes to the policy debate.

Acknowledgments

We are deeply grateful for the support we have received from our sponsors throughout this project. The analysis reported here required the development of new methodologies, software, and data, and we encountered delays and obstacles along the way. If it were not for the patience and farsightedness of our sponsors, this project could never have been completed. Their guidance and funding enabled us to provide an analysis of compensation alternatives from the viewpoint of both the active and reserve components, and we believe that the analytical apparatus can be fruitfully extended and applied to other challenging issues of compensation, retention, and force shaping. With greatest pleasure, we acknowledge our debt of gratitude to Brigadier General (ret.) Denny Eakle, director of the 10th QRMC; Saul Pleeter, research director of the 10th QRMC; and David S. C. Chu, Under Secretary of Defense, Personnel and Readiness. We are equally pleased to recognize the support received from John Winkler, Principal Deputy Assistant Secretary of Defense for Reserve Affairs, who, early on, provided support and funding that allowed us to develop a precursor to the current model for an analysis of reserve retirement reform. Finally, we could not have conducted our empirical analysis without data provided by the Defense Manpower Data Center, a remarkable organization that has become a true national security resource.

At RAND, we are grateful to Daniel Clendenning for much of the initial model programming, and we appreciate how much our work has benefited from the comments of our reviewers, Al Robbert and Jacob Klerman, who deserve our thanks for their thorough, constructive reviews. We also thank the many individuals at RAND and on the working group of the 10th QRMC for their comments and observations while this work was in progress.

Abbreviations

ABP	annual basic pay
ACOL	annualized cost of leaving
BAH	basic allowance for housing
BAS	basic allowance for subsistence
COLA	cost-of-living allowance
CPI	Consumer Price Index
DACMC	Defense Advisory Committee on Military Compensation
DB	defined benefit
DC	defined contribution
DMDC	Defense Manpower Data Center
DoD	U.S. Department of Defense
ERISA	Employee Retirement Income Security Act of 1974
FERS	Federal Employees Retirement System
IT	information technology
MBP	monthly basic pay
PDV	present discounted value

QRMC   Quadrennial Review of Military Compensation

RMC   regular military compensation

S&I   special and incentive

WEX   Work Experience File

YOS   year of service

CHAPTER ONE

Introduction

The current U.S. military retirement system dates back to the post–World War II era, when a common system was defined for both officer and enlisted personnel. The current system vests active-duty members at year of service (YOS) 20 with immediate benefits. Yet, numerous studies and commissions have criticized the system and asked whether a system developed following WWII is the best system for a modern military.

In the immediate post-WWII era, one of the concerns was to prevent the personnel force from becoming top-heavy with senior personnel. The vast majority of the millions of individuals who served in the armed forces in WWII were compelled to leave the force at the end of the war. However, many senior personnel who, in the absence of the war, might have been expected to leave the military instead remained. While up-or-out policies or involuntary separation could have addressed the issues of superannuation and clogged promotion opportunities, such approaches to large-scale force downsizing could have undesirable political ramifications, as was demonstrated during the post–Cold War downsizing of the early 1990s, when involuntary separation and large-scale reductions in accessions were initially used to reduce the force, resulting in opposition from the services, service members, and veterans' groups. The retirement reform of 1948 offered generous retirement benefits to personnel with 20 or more years of service as an inducement to leave service. This system proved successful in preventing an excess of senior personnel while providing a financial benefit that smoothed the transition to civilian life and quieted political opposition to the large-scale separation of personnel. The retirement system enabled personnel to exit the military voluntarily, and the same principle is relevant today with the all-volunteer force. Further, viewed in a dynamic sense, it ensured that senior positions would be voluntarily vacated through retirements, which meant that promotion opportunities for junior personnel would be sustained. Promising future leaders therefore would not be faced with dead-end military careers and would have the opportunity to reach top positions.

Despite its success in rebalancing the force's experience mix after WWII and maintaining advancement opportunities, the retirement-benefit system resulted in other outcomes that were not as desirable. In today's world, the main criticisms are that it results in a military compensation system that is excessively costly, is inequitable for members who do not serve long enough to reach the 20-year vesting point, lacks comparability with the civilian sector because it does not provide an employer-funded 401(k) plan, and hampers force-management flexibility by encouraging career lengths that may be too short or too long for some career fields, even if the overall result of a 20-year career is desirable. Regarding cost, the compensation system is considered inefficient because it back-loads military compensation in deferred compensation. The typical service member discounts deferred benefits at rates much higher than the rate at which the government discounts future costs, as we discuss later. Because of this high rate, it is more costly to increase retention by increasing deferred compensation than by increasing current compensation. Conversely, it is less costly to increase retention by increasing current compensation than by increasing deferred compensation. The military compensation system would be more efficient in terms of achieving a given force at a lower cost if a higher share were in current rather than deferred compensation. We discuss the case for reform in Chapter Two.

While much of the debate surrounding military retirement reform has focused on the active-duty system, the reserve retirement system has also been subject to criticism. The reserve system differs from the active system in several ways, but the key difference is that a reservist who achieves 20 creditable years of service must wait until age 60 before he or she can begin receiving benefits, unlike his or her active-duty counterpart, who can begin receiving benefits as soon as YOS 20. Thus, a regular Army member, for example, with 20 years of service could retire at age 40 and receive benefits, while an Army reserve member with 20 years of creditable service must wait until age 60 for benefits. With reserve forces being used to a greater extent than previously in military operations, questions are being raised about the adequacy of the reserve retirement benefit and, specifically, differences in the age at which active and reserve members can begin claiming benefits.[1] Congress has considered several bills to reduce the age at which reservists can begin receiving their retirement benefits.

More broadly, numerous proposals have been offered to address the criticisms of the active and reserve retirement systems. Most recently, the Defense Advisory Committee on Military Compensation (DACMC) recommended a three-part reform (see DoD, 2006). The features include (1) a 401(k)-like plan to which the government would contribute in the range of 5 percent of basic pay and that would vest after YOS 10; (2) a defined benefit (DB) plan that would pay an annuity beginning at age 60, vesting after 10 years of service, using a formula similar to the current retirement annuity formula; and (3) additional current compensation to achieve force-management goals that could come in various forms, including separation or transition pay of limited duration for those who leave after the vesting point, increases in basic pay, bonuses, or gate pay that would be paid to those completing key career milestones, such as achieving 10, 15, 20, 25, and 30 years of service. The DACMC provided evidence to suggest that this structure would be more efficient in terms of producing similar retention at lower cost, more equitable in terms of allowing more members to become vested, and more flexible in terms of force management because of the ability to target gate pay and separation pay to achieve desired retention profiles.

---

[1] The January 2008 report of the Commission of the National Guard and Reserves (DoD, 2008a) provides an excellent summary of changes in the roles of the reserve components. Asch, Hosek, and Loughran (2006) discuss the differences between the active- and reserve-component retirement systems and the implications for equity, force management, and cost.

Assessing the efficiency, equity, and flexibility of the current and alternative compensation systems requires a model that recognizes the career decision processes of individual service members, the heterogeneity of their preferences, the uncertainty of the environment in which they make career decisions, the time path of these decisions, and the organizational structure and policy context in which they make these decisions. Furthermore, given the greater operational role of the reserves and the importance of total force compensation and personnel policy, the model must consider both active and reserve career decisions.

The model best suited for such an assessment is a stochastic dynamic programming model, which we describe in Chapter Three. The dynamic programming approach is well suited to analyzing compensation-reform proposals because its parameters can be estimated from data on active and reserve retention under current compensation, and the estimated model can then be used to simulate compensation proposals. These capabilities are highly valuable because there have been no major changes in the military retirement system, so no actual data exist on the effects on retention behavior of major variations in the retirement system. The dynamic programming model is formulated in terms of the parameters that underlie the retention decisionmaking process. Because the dynamic programming model is based on the factors underlying the decision process, rather than on a specific compensation system and retention outcomes, the model permits assessments of alternative compensation systems.

In the context of the retention of military officers, such a model was first formulated and estimated by Gotz and McCall (1984) at RAND and called the *dynamic retention model*. Estimating the model parameters with data, using such methods as maximum likelihood, is computationally complex and proved difficult given the computer technology available in the 1980s and 1990s; however, Daula and Moffitt (1991) estimated the model with data from two enlisted Army cohorts. The Gotz-McCall model was extended in several ways in the 1990s. Asch and Warner (1994b) incorporated performance into the model, and Asch and Warner (1994a) calibrated a simulation model that enabled them to estimate the steady-state retention, performance incentives, and cost implications of alternative military compensation and retirement policy alternatives. Asch, Johnson, and Warner (1998) extended the simulation model further to estimate the retention, cost, and productivity effects of transitioning to a military retirement system that resembled the Federal Employees Retirement System (FERS). An updated version of the Asch-Warner simulation model was used to assess retirement alternatives for the DACMC, described earlier. Asch and Hosek (1999) employed the simulation model to analyze the behavioral and cost implications of the Triad proposal, as well as other military compensation and retirement-reform proposals, that addressed concerns among the military leadership in the late 1990s (when the services struggled to meet their military recruiting and retention targets) about the adverse effects on retention and morale due to the reduced value of retirement benefits under the reform plan implemented in 1986 (often referred to as *REDUX*). Hosek et al. (2004) incorporated the enlistment decision into the dynamic retention model framework in a study of the recruitment and retention of information technology (IT) personnel. They also modeled skill accumulation—the learning of IT skills through training and experience provided by the military—with the assumption that the skills are transferable and so increase the civilian opportunity wage. They included a switching cost that is imposed if the individual breaches his or her military contract by leaving before the end of the term. In calibrating their model, they estimated the distribution of the preference for military service in the youth population. They also analyzed the attractiveness of military IT occupations (compared to non-IT occupations), where IT occupations (by providing valuable, transferable training) provide a pathway to high-paying civilian jobs when the member leaves the military. Mattock and Arkes (2007) adapted the Gotz-McCall model to analyze incentive pay for Air Force officers, including a provision requiring a multiyear commitment.

Advances in computer hardware and software make estimation of the dynamic retention model feasible. In this monograph, we estimate a dynamic retention model of active and reserve retention using data provided by the Defense Manpower Data Center. The data are drawn from the Work Experience File (WEX), which tracks the careers of active and reserve personnel. We supplement the data with pay data and other necessary model inputs. The model is estimated for each service branch, and we use the parameter estimates to conduct policy simulations for the enlisted force.

In addition to estimation and policy simulation based on the active/reserve dynamic retention model, we undertake separate calculations to compute the actuarial value of members' wealth at different possible separation points under the current military compensation system and under the proposed alternatives. These computations provide information on how the alternatives change members' assets and, specifically, whether they make members better off in terms of wealth. We compute the actuarial values under alternative assumptions regarding the personal discount rate to assess how sensitive the values are to alternative assumptions about the discount rate.

The 10th Quadrennial Review of Military Compensation (QRMC) builds on the groundwork laid by the DACMC by developing specific retirement proposals that address the criticisms raised by past studies and commissions. The chief purpose of this monograph is to apply our model to provide policy analysis of these proposals. The results of our policy analyses provide an indication of the promise of the QRMC proposal in terms of its force-management effects and how member wealth changes in terms of the actuarial value of compensation. If the simulations demonstrate that these proposals have potential, the next step would be a closer study, perhaps in the context of a pilot test.

The following chapters discuss the case for retirement reform, describe our analytical approach and estimated model, present the results of various policy simulations, and offer conclusions. Chapter Two discusses the case for compensation reform in more detail, while Chapter Three describes our analytical approach. Chapter Four presents the model estimates and simulations of the current compensation system. Chapter Five describes the policy alternatives and presents simulations of these alternatives. Our conclusions are presented in Chapter Six.

The Case for Retirement Reform

To evaluate the effectiveness of the current compensation and personnel management systems, one must state the goals of the systems. From a force-management perspective, these systems should attract and retain the quantity and quality of required personnel; provide training and develop personnel so that they have the skills necessary to be productive; provide them with incentives to perform well and pursue activities that demonstrate and develop their capabilities; induce them to seek positions in which those capabilities are put to their best use, including higher-ranked positions; and separate personnel voluntarily at some point when it is best for the organization. As part of this process, these systems should recognize arduous and hazardous duties that are far from home. Furthermore, they must recognize unusual elements of the military personnel system, such as the hierarchical organizational structure, in which promotions feed the upper ranks, and the virtual lack of lateral entry from the civilian sector.[1]

The current compensation system, consisting of the basic-pay table, various allowances, special and incentive pays, retirement pay, a federal tax advantage, and other benefits has been, by and large, quite successful in meeting the manpower needs of the all-volunteer force (Bicksler, Gilroy, and Warner, 2004; Asch and Hosek, 2004). The compensation system has been stable over time, and the common pay

---

[1] The goals of the military compensation and personnel system have been articulated in a variety of studies. See, for example, Asch and Warner (1994b), the Report of the Seventh QRMC (DoD, 1992), and the Report of the Defense Advisory Committee on Military Compensation (DoD, 2006).

table across occupational areas underscores the notion of equity: Different members in different services are equally compensated, given their years of service and rank. The pay table provides returns on advancement by structuring pay such that the return on promotion is greater than the return on another year at the same grade. Furthermore, the ranks provide explicit rungs on a career ladder, with those who move up having the opportunity to receive higher pay. Given that promotion is differential in terms of performance, these ranks provide an incentive for members to exert effort and demonstrate their skills and talents.

The various special and incentive pays have different rationales. Bonuses, for example, enable the recruitment and retention of personnel in critical skill areas. The retirement system creates a strong incentive for military personnel to stay beyond 10 years and to leave after 20 years. The retention of mid-career personnel provides a return on investment in training during the first and second terms and creates a pool of experienced personnel from which senior leaders can be drawn. Most importantly, a key role of the retirement system is inducing members to separate voluntarily. As discussed by Warner (2006), the 1948 Advisory Commission on Service Pay (the Hook Commission) found that the current system was well suited to preventing a superannuation of the force and ensuring youth and vigor. In the absence of the retirement system, the military would have to use involuntary separation, which would hurt morale, possibly adversely affecting productivity.

Many commissions and study groups have examined the compensation system; Christian (2006) provides a review of these studies. The main criticisms of the system are that it is unfair and excessively costly and that it inhibits force-management flexibility.

The system is considered unfair because private-sector pension systems are governed by the Employee Retirement Income Security Act of 1974 (ERISA) (P.L. 93-406), which requires vesting no later than after five years of tenure with a company (or seven years under graduated vesting). Because service members are not vested until they reach 20 years of service, the current system is considered unfair to junior personnel. The system is considered incomparable with what is available in the private sector. Furthermore, it lacks comparability with the federal civilian sector, because the military retirement system is a DB plan, in which the benefit is defined by a formula, and does not include a defined contribution (DC) plan funded by government contributions.

Those who view the system as excessively costly focus on the retirement annuity paid to members who are still relatively young and in the civilian workforce, working on their second career. The President's Commission on Military Compensation (also known as the Zwick Commission) recommended converting the military retirement system to one providing an old-age annuity and a trust fund from which members separating from the military could withdraw their entire portion. The commission argued that the change would bring 35- to 40-percent savings in retirement cost and would reduce the incentive to exit upon completing 20 years of service (Cooper, 1978). In 1985, the President's Private-Sector Survey on Cost Control (the Grace Commission) recommended eliminating these annuity payments and initiating a payout of the retirement benefit to active-duty members beginning at age 60 or 62. Asch and Warner (1994a, 1994b) found that the Army could achieve the same retention patterns as under the current system, but at less cost, by increasing basic pay across the board. Such a front-loaded compensation system reduces cost, because members discount future benefits at a higher rate than the government discounts future costs. A dollar paid today is worth a dollar to the service member and costs a dollar to the government. But a dollar paid 10 years from now is worth less to the service member today than the amount the government must implicitly invest today to have a dollar to pay in 10 years. Because each dollar the government expends today is worth more to the service member the sooner it is paid relative to its cost to the government, front-loaded compensation is more cost-effective. To achieve a given level of retention, fewer dollars must be expended if pay occurs in the form of basic pay than in the form of retirement. Asch and Warner estimated that costs would be about 5-percent lower in the steady state under this more front-loaded system.

Many recent studies have focused on the issue of personnel management flexibility. As the United States has moved from the Cold War–era construct of the large, standing military toward a more modular and expeditionary force structure that is more easily tailored to diverse scenarios small and large, the advantages of flexibility in managing personnel have become paramount. The current compensation system has produced a stable flow of personnel with highly similar career paths, and the shape of career paths has been heavily influenced by the retirement-benefit system. Yet these career paths may be too constrained. The length of a career should also depend on the productivity of military personnel with respect to their experience and grade and the cost of training, developing, and retaining personnel, given their external opportunities. Unanimously, the various studies concluded that, given the myriad of skills required in the military (those that require youth and vigor and those that do not), a retirement system that can accommodate shorter and longer careers would be desirable.

Also of concern from a management flexibility perspective is whether personnel, especially officers and noncommissioned officers, spend the right amount of time in assignments. Longer time in an assignment would allow more time to learn a job and to capture the returns of greater job experience. Longer assignments are more feasible if longer careers are possible. Furthermore, as discussed by Rostker (2005), Schirmer et al. (2006), and Warner (2006), short assignments mean that members are often rotated before they see the results of their efforts, and this can give rise to perverse incentives, as members may pursue short-term goals. Another question raised is whether members might stay longer in a given grade, rather than being forced out by up-or-out rules. Staying longer within a grade would permit skill specialization and enable members to be productive in specific tasks without the requirement of being moved to a supervisory position or forced out by high-year-of-tenure rules. More variation in time in grade would mean more variable career tracks for some personnel.[2] The current compensation system does not permit much variation in pay by YOS or in career lengths, as we show next.

---

[2]   It should be noted that the model assessment of retirement alternatives includes the effects on the grade distribution of personnel (though these results are not shown in Chapter Four), but it does not consider the feedback effects of changes in retention on grade distribution and time in grade.

Cash compensation for military personnel can be divided into regular military compensation (RMC), special and incentive (S&I) pays, bonuses, and miscellaneous allowances and cost-of-living allowances (COLAs). RMC is the sum of basic pay, housing allowance, subsistence allowance, and the federal tax advantage owing to the nontaxability of the allowances. Average cash compensation in 2004 was around $44,000 for enlisted personnel (see Table 2.1), and RMC accounted for about 90 percent of that amount. S&I pays averaged $380 to $1,750 for enlisted personnel in 2004. These averages might seem low, but the averages are taken across all personnel, and most personnel do not receive any given S&I pay. Also, many S&I pays are not large. For instance, the average amount of proficiency pay for airmen who received it was $2,373, but only 6 percent received it. The same was true of bonuses, miscellaneous allowances, and COLAs.

With respect to military careers, average cash compensation in 2004 rose for enlisted personnel, from around $30,000 at entry to about $75,000 at the 30th year, an increase of $1,500 per year (see Figure 2.1). Although the services share a common pay table, and longevity increases are automatic, the promotion system can create pay differences among personnel in different occupations. Promotion speeds of enlisted personnel vary across the services and have been

Table 2.1
Average Enlisted Pay, 2004

Category of Cash Compensation	Service ($)			
	Army	Air Force	Marine Corps	Navy
RMC	40,784	41,854	37,764	41,091
S&I pays	1,048	383	981	1,746
Bonuses	428	929	414	1,381
Miscellaneous allowances and COLAs	2,065	1,818	1,306	1,722
Total	44,329	44,981	40,463	45,937

SOURCE: Authors' tabulations based on military pay files. Amounts are rounded.

Figure 2.1
Average Total Enlisted Pay, by Service and YOS, 2004

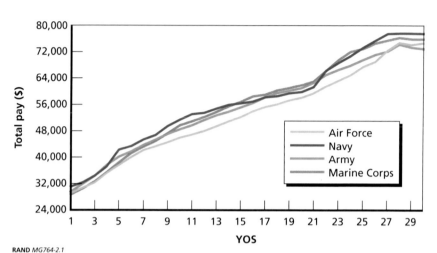

RAND MG764-2.1

fairly stable over time for each service (Hosek et al., 2004), with the
Air Force having the slowest promotion time to pay grade E-5 and
the Marine Corps having the fastest, which largely accounts for the
Air Force having lower average pay than the other services after YOS
6. The comparison of average cash compensation by service over the
course of a career suggests that, on average, pay differences are not
large. Variation in enlisted pay within a service comes mainly from
S&I pays and bonuses and secondarily from differences in promotion
speed (Asch, Hosek, and Martin, 2002).

An examination of the years of service or experience mix of personnel across occupational areas within a service suggests that the service branches have generally relied on S&I pays and bonuses to generate similar career lengths across occupational specialties (Asch, Hosek,
and Martin, 2002). That is, the variation in these pays has resulted in
similar career lengths across the force. Consequently, the experience
mix of the career force, particularly after the first five years of service,
is quite similar across career fields. The greater "front-end" variation
in the YOS 1–5 category probably reflects differences in attrition rates
(i.e., leaving before completing the first term), enlistment bonuses, and
adjustments in recruiting targets driven by unexpectedly high or low

retention in higher years of service. Although the experience mix is similar across occupations within a service, there are some differences across the services. These differences arise from the services' roles and missions and the inherent attractiveness of the training, career tracks, living environments, and opportunities for deployment.

The similarity in experience mix across occupational areas within a service suggests that variations in compensation, including promotion policy and the use of S&I pays across occupational areas within a service, are used to achieve a similar career length. As a result, determining the manpower requirements for a military activity is conditional on the expected flow of personnel by YOS. That is, the compensation and personnel systems operate to provide a supply of personnel, and the manpower system makes allocations subject to supply constraints. In an alternative system, manpower requirements would emerge after an assessment of the productivity and cost of different manpower configurations, and the compensation system would be sufficiently flexible to ensure that the optimal requirement would be met.

To the extent that the required experience mix by and large reflects the compensation structure but the experience mix produces careers that are too uniform, too short, or too long for some assignments and careers, a change in the structure of compensation is needed to permit greater personnel management flexibility.

The various studies and commissions recommended specific proposals. Most recently, the DACMC proposed a three-part system that includes (1) a 401(k)-like plan to which the government would contribute in the range of 5 percent of basic pay and the contributions would vest after YOS 10; (2) a DB plan that would pay an annuity beginning at age 60, vest after YOS 10, and use a formula similar to the current retirement annuity formula; and (3) additional current compensation to achieve force-management goals that could come in various forms, including separation or transition pay of limited duration for those who leave after the vesting point, increases in basic pay or bonuses, or gate pay, that would be paid to those completing, say, 10, 15, 20, 25, or 30 years of service.

The DACMC-proposed structure has a number of attractive features. More members would vest because vesting occurs earlier, and, therefore, the new system would be more equitable to junior personnel. It could facilitate different career lengths by occupational area to the extent that the transition or separation pays would be targeted by occupation, and the amount and timing of payment could vary. To ensure retention and to shape the retention profile, the third component, career gate pay, could be increased at various points in the career. Doing so would result in a less back-loaded compensation system and, therefore, a more efficient system in the steady state. A final advantage is that the new system could integrate the active and reserve retirement systems under the same plan.

Although the DACMC considered some specific plans that had the three components it recommended, it did not recommend a specific proposal. The DACMC assessed these specific plans using an updated version of the calibrated simulation model developed by Asch and Warner (1994a) for the Army enlisted force. The model parameters are not estimated but are calibrated, and the dynamic retention model focuses on active-duty retention decisions. With calibration, parameters are found, via trial and error, so that the predicted grade and experience force profiles from the model match the Army's actual profiles, and to make the prediction, the model relies on a small set of hypothetical individuals who vary in terms of taste for service. With estimation, as in this monograph, actual data on approximately 30,000 military personnel are used, and model parameters are found by maximizing the likelihood that the retention behavior predicted by the model data fits the behavior of the actual data.

In this monograph, we consider the DACMC proposals, as well as another alternative proposed by the QRMC, to address the criticisms of the current system. We estimate our dynamic retention model of active and reserve behavior for each service branch and employ it to analyze the compensation proposals. The model and estimates are discussed in the next two chapters, followed by an evaluation of the proposals and a concluding chapter.

CHAPTER THREE

Analytic Framework

An analysis of military compensation reforms requires a theoretical framework that can describe retention behavior in response to complex changes in military compensation. It is important to consider not only retention in the active components but also participation and retention in the selected reserve. The growing prominence of the reserves in the U.S. defense system is characterized by the shift from their role as a strategic reserve to that of an operational reserve. The active-duty force is the primary source of experienced, skilled reservists, so it makes sense to consider active and reserve personnel within a common analytic framework. Policies that affect the active-duty force may have repercussions for the reserve force, for example. As another point, it is valuable to ground the analysis of compensation policy in actual behavior, if possible. Parameters estimated from data on active retention and reserve participation and retention are preferable to best guesses about these parameters. Finally, because there have been no major changes in the retirement system and, consequently, no data on retention following such changes, an approach is needed that allows for the analysis of major compensation reforms without relying on the existence of prior variations in such reforms.

Our approach satisfies these criteria. We use a dynamic programming model of active-duty retention, affiliation with the reserves after active duty, and reserve participation. We estimate the model with longitudinal data on active and reserve service, and these data are augmented with additional data, as described later in this chapter. Packaged software is not available for dynamic programming models, so we have written software for estimation and simulation.

This chapter describes our model informally and then formally, and we further discuss data and estimation in Appendix B. This chapter also outlines the use of the model for policy simulation. The next chapter presents the parameter estimates and provides information on how well the model fits the data. Some readers may wish to skip to the next chapter once they have read the informal description of the model.

Overview of the Model

The purpose of the model is to lend insight into the effect of changes in military compensation on a service member's willingness to continue on active duty or, if leaving active duty, to participate in the selected reserve. The complexity of analyzing the effect of military compensation comes not only from the fact that it involves both current and deferred compensation, but also from the connection between military service in the current period and military opportunity and compensation in future periods. For example, retirement benefits, a major form of deferred compensation, can affect current retention, and current retention can affect progress toward promotion and the eligibility for and amount of retirement benefits. Active and reserve retention is also affected by opportunities. An active-duty service member must consider in each period whether it makes sense to continue on active duty or to become a civilian or selected reservist, and an ex-active-duty service member must consider whether to serve in the selected reserve or simply be a civilian. Two factors further complicate the analysis of military compensation: namely, individual differences in terms of preference for serving on active duty or serving in the reserves and the role of unanticipated factors—shocks—that can affect the appeal of any of the alternatives. This discussion could well be extended to include many other factors that may affect an individual's retention behavior, e.g., marital status, spouse employment and earnings, educational aspirations, health conditions, housing, locale, training, equipment, leadership, the opportunity to deploy. But with respect to our work, which addresses the structure of military compensation, we focus on a more limited but still challenging set of factors: current or deferred; active, reserve, or civilian; individual preference for active or reserve service; and shocks, which capture the influence of some of the additional variables, though indirectly.

Figure 3.1 describes the possible decision nodes for an individual who begins service on active duty. These are the decision points embedded in our model in the sense that, in each period, an individual must consider all opportunities. The figure includes three periods, but our model considers a 40-year work life beginning with active duty. In the model, individuals are assumed to maximize their lifetime utility, which depends on their earnings, their preference for serving on active duty or in the reserves relative to being a civilian, and the implicit value of other factors (shocks) that affect one's satisfaction, or lack of satisfaction, with current conditions. The value of an opportunity in the current period, such as continuing on active duty, depends on the current compensation it offers, plus the monetary value that the individual

Figure 3.1
Sample of an Active-Duty Service Member's Possible Decision Points

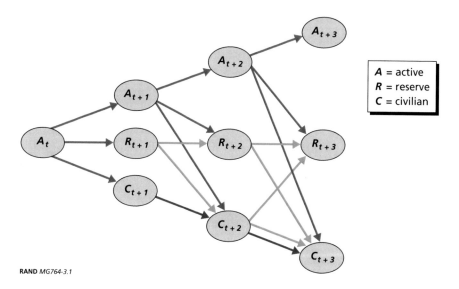

A = active
R = reserve
C = civilian

assigns to his or her preference for that opportunity (i.e., the intrinsic value to the individual of serving on active duty), plus the value of shock, which, as mentioned, may represent many aspects of military or civilian positions. The value of an opportunity in the current period also depends on its effect on the value of the opportunities available in the next period.

The diagram shows the service member's possible decision points, and it also illustrates two assumptions that we have built into the model to simplify the task of estimating it. The diagram starts with an active-duty service member at period t. In prior periods, he or she has been on active duty. At t, the service member chooses between continuing on active duty, transitioning to the selected reserve, or becoming a civilian. The model assumes that a service member who leaves active duty cannot reenter it. The diagram illustrates this idea as follows: The red arrows indicate leaving active duty, and there are no arrows to indicate a return to active duty. Further, we assume that, if the service member leaves active duty and joins the selected reserve, in future periods, he or she may move back and forth between reserve and civilian status (indicated by the green arrows). But if he or she leaves active duty to become a civilian, he or she remains a civilian in future periods (purple arrows). The second simplifying assumption, then, is that only if the service member joins the selected reserve in the period immediately following active duty may he or she serve in the reserves in future periods. These two assumptions reduce the number of possible decisions and, importantly, are consistent with reality. Very few of those who leave active duty ever reenter active duty, and most of the reentry that does occur is in the first few years of active duty. Also, about 92 percent of ex-active-duty service members who ever join the selected reserve do so within one year of leaving active duty, and about 95 percent do so within two years. When we estimate the model, we define the period length to be two years. These simplifying assumptions reduce the computational burden of estimating the model, but they could be relaxed in the future if reentry into active duty from the selected reserve became more frequent.

The model is estimated on data for enlisted personnel entering active duty between 1990 and 2007 and followed over time, to the extent permitted within our data window. Many leave active duty after the first or second term of enlistment, while others continue. Some leavers join the selected reserve, and their participation (e.g., continuation, exit, reentry) in the reserves is followed. The model assumes that individuals optimize over time and that they compare the utility of the alternatives available in each period, given their decision point. This depends on current and deferred military compensation and on civilian earnings, and the values of these are known from military pay tables and from civilian earnings data. Utility also depends on elements that are not observed—namely, individual preferences for active and reserve service, the values of the shocks for each alternative in each period, and the personal discount rate. The process of estimating the model is the process of finding the parameters of the preference and shock distributions, as well as the discount rate, that fit the data best. The formal model and estimation method are discussed in the following sections.

Once we have the estimated parameters, we use them to simulate alternative compensation policies. We create a population of synthetic individuals whose preferences for active and reserve service are drawn from the preference distribution and whose shocks are drawn from the shock distribution. The new compensation policy replaces the old, and as a result, the current and deferred military compensation changes and creates new incentives to stay in or leave active duty or the selected reserve. The simulation shows the choices that each synthetic individual makes, and we use the results to compute the retention profile, costs, and man-years of the policy proposals versus the current policy.

Active and Reserve Retention Model

Gotz and McCall (1984) were the first to apply dynamic programming to military retention. They analyzed the stay-or-leave decisions of officers in the active component of the Air Force, and we extend their model to include the possibility of participation in the reserve component after serving in the active component. Other work applying dynamic programming to military retention includes Daula and

Moffitt (1991), Asch and Warner (1994a), Asch, Johnson, and Warner (1998), Hosek et al. (2004), and Mattock and Arkes (2007). These models focus exclusively on retention in the active component.

Our model assumes that an individual begins his or her military career in an active component, and, in each period, the individual can choose to continue on active duty, leave the military to hold a job as a civilian, or leave the military to join a reserve component and hold a job as a civilian. Individuals are assumed to differ in their preferences for serving in the military. Each individual is assumed to have given, if unobserved, preferences for active and reserve service that do not change. The individual has knowledge of military pay and retirement benefits, as well as civilian compensation. In each period, there are random shocks associated with each of the alternatives, and the shocks affect the value of the alternative. As shown next, the model explicitly accounts for individual preferences and military and civilian compensation, and in the context, shocks represent current-period conditions that affect the value of being on active duty, being in the selected reserve, or being a civilian. Examples of what may contribute to a shock are a good assignment; a dangerous mission; an excellent leader; inadequate training or equipment for the tasks at hand; a strong or weak civilian job market; an opportunity for on-the-job training or promotion; the choice of location; a change in marital status, dependency status, or health status; the prospect of deployment or deployment itself; or a change in school tuition rates. These factors may affect the relative payoff of being in an active component, being in a reserve component, or being a civilian. The individual is assumed to know the distributions that generate the shocks, as well as the shock realizations, in the current period but not in future periods.

Depending on the alternative chosen, the individual receives the pay associated with serving in an active component, working as a civilian, or serving in a reserve component and working as a civilian. In addition, the individual receives the intrinsic monetary equivalent of the preference for serving in an active component or serving in a reserve component. These values are assumed to be relative to that of working as a civilian, which is set at zero. In considering each alternative, the individual takes into account his or her current state and type. *State* is defined by whether the individual is active, reserve, or civilian and by the individual's active years of service, reserve years of service, total years (age), pay grade, and random shocks. *Type* refers to the level of the individual's preferences for active and reserve service. The individual recognizes that today's choice affects military and civilian compensation in future periods. Although the individual does not know when future military promotions will occur, he or she does know the promotion policy and can form an expectation of military pay in future periods. Further, the individual does not know what the realizations of the random shocks will be in future periods. The expected value of the shock in each state is zero. Depending on the values of the shocks in a future period, any of the alternatives—active, reserve, or civilian—might be the best at the time. Once a future period has been reached and the shocks are realized, the individual can reoptimize (i.e., choose the alternative with the maximum value at that time). The possibility of reoptimizing is a key feature of dynamic programming models that distinguishes them from other dynamic models. In the current period, with future realizations unknown, the best the individual can do is estimate the expected value of the best choice in the next period, i.e., the expected value of the maximum. Logically, this will also be true in the next period, and the one after it, and so forth, so the model is forward-looking and rationally handles future uncertainty. Moreover, the model presumes that the individual can reoptimize in each future period, depending on the state and shocks in that period. Thus, today's decision takes into account the possibility of future career changes and assumes that future decisions will also be optimizing.

The general structure of the model is as follows:

$$Y_{jk}(s_t;\gamma) = w_{kt} + \gamma_k + \beta \, \text{Emax}\left(Y_{ka}(s_{t+1};\gamma), Y_{kr}(s_{t+1};\gamma), Y_{kc}(s_{t+1};\gamma)\right)$$
$$+ \, \varepsilon_{kt}, \tag{1}$$

where

Y_{jk} = value function for transition from j to k

$j,k \in \{\text{active, reserve, civilian}\}$

$s_t = s_t(ay_t, ry_t, t, g_t)$ state in period t

ay = active years

ry = reserve years

t = total years

g = pay grade

w = current pay

$$\gamma_k = \begin{cases} \gamma_a & \text{monetary value of preference for serving in active component} \\ \gamma_r & \text{monetary value of preference for serving in reserve component} \\ 0 & \text{preference for civilian job} \end{cases}$$

β = personal discount factor

Emax = expected value of the maximum

ε_{kt} = random shock.

Given current activity j, the value function for transitioning to activity k is Y_{jk}, where j and k represent active, reserve, and civilian. The value of the value function depends on current pay, the monetary value of the preference, the present value of being able to choose the best alternative in the next period, and the random shock. Current pay depends on the state. For example, military pay in an active component depends on active years of service and pay grade, and military pay in a reserve component depends on active and reserve years of service and pay grade. Civilian pay depends on total years. In addition, civilian pay includes the military retirement-benefit payment if and when the individual is eligible to receive it. Under the current military retirement-benefit system, a service member with 20 years in an active component is eligible for an active-duty retirement benefit payable immediately upon leaving the active component and becoming a civilian. A reservist with 20 years of creditable service is eligible for a reserve retirement benefit, which, during the period of our data, began at age 60. Active and reserve retirement-benefit amounts depend on active years of service and on active plus reserve years of service, respectively, and on pay grade—in particular, the highest three years of military pay.

The term γ_k is the monetary value of the individual's preference for the activity, i.e., γ_a for active service and γ_r for reserve service. The personal discount factor, β, is defined as $1/1+r$, where r is the personal discount rate. The operator Emax takes the expected value of the *maximum* of the value functions in the next period. Intuitively, from the perspective of the current period, the Emax expression represents the fact that the individual can reoptimize in the next period once the random shocks in that period have been realized. Because future shocks are not known in the current period, the individual computes the expected value, given that the best choice will be taken. The term ε_{kt} is the random shock in activity k in period t.

The model is structured as a Markov process. The current state is assumed to capture all relevant information from the individual's history. For instance, for a service member who has served continuously in an active component, the state records the number of years of active duty and the current pay grade, and the exact timing of past promotions does not figure into the state. For this individual, reserve years are zero, and total years coincide with active years. The individual's optimal decision is characterized by knowing that he or she will make the best decision in all future periods, conditional on the information available in those periods. Using this insight, the model can also be written as follows:

$$Y_{jk}\left(s_t;\gamma\right) = w_{kt} + \gamma_k + \beta \sum_m \pi_{km}\left(s_{t+1} \mid s_t;\gamma\right) Y_{km}\left(s_{t+1};\gamma\right) + \varepsilon_{kt}$$

$$\pi_{km}\left(s_{t+1} \mid s_t;\gamma\right) = \text{probability alternative } m \text{ is max}\left(Y_{km}\left(s_{t+1};\gamma\right)\right),$$

$$m \in \{a,r,c\}. \tag{2}$$

The probability $\pi_{km}\left(s_{t+1} \mid s_t;\gamma\right)$ is the probability that alternative m is the best choice, i.e., has the highest value. The subscript km indicates a transition from alternative k, the alternative chosen in period t, to alternative m in period $t+1$. Our approach to estimating the model relies on these optimal transition probabilities. As we show, they can be defined as functions of the value functions, and the transition probabilities for an individual's choice in each period can be multiplied together to obtain the probability of the individual's actual sequence of choices as shown in the data on the individual's active and reserve careers. Thus, the transition probabilities are a function of the value functions in each period, and, in turn, they are functions of the underlying parameters of the model. This is discussed further in the section on estimation in Chapter Four.

We imposed several constraints in applying this model. First, once a person leaves the active component to become a civilian or a reservist, the person may not reenter the active component. Second, in the period in which a person leaves the active component, he or she may choose to be either a civilian or a reservist. But if the person chooses to be a civilian, he or she may not join the reserves in later periods. In contrast, if the person chooses to be a reservist, he or she may move back and forth between reserve and civilian in future periods if he or she chooses to do so. Third, we assume that individuals enter active duty at age 20 and have a work life of 40 years, and we specify a period length of two years. Fourth, we assume that the individual begins service on active duty.

The first constraint prohibits reentry into an active component. Although some reentry occurs, it is quite limited and typically takes place in the first few years of service as so-called prior-service accessions join the force. Imposing the constraint reduces the state space with, we feel, little impact on our parameter estimates and policy simulations but has the advantage of reducing computation time. If the movement from reserve to active duty becomes more frequent in future years, the model can readily be altered to accommodate this.

The second constraint (i.e., to join the reserves in the first period after active duty or not at all) is consistent with the data. The WEX data indicate that more than 94 percent of those who ever join the selected reserve after serving in an active component do so within the first two years after active duty (see Table 3.1). Therefore, this constraint appears to be a defensible simplification. In fact, nearly 90 percent join within six months after active duty, and more than two-thirds

Table 3.1
Among Those Ever Joining the Selected Reserve After Service in an Active
Component, Percent Joining Within Three Years

Service	Percent Joining				
	One Month	Six Months	One Year	Two Years	Three Years
Army	73	88	91	94	96
Navy	67	88	93	95	97
Air Force	86	92	93	95	96
Marine Corps	81	89	92	94	96

appear to join almost immediately. To incorporate this constraint and the first, we assign members in our data who join the selected reserve after two years or who return to active duty a zero weight in the likelihood function.

Like the first two constraints, the third constraint (a period length of two years) also reduces the size of the state space and reduces computation time. The fourth constraint focuses the analysis on the total force from the perspective of the active component and its contribution to the selected reserve. This accounts for a large portion of the personnel serving in the active and reserve components but excludes individuals who join the reserves without prior active experience; they deserve attention in future work.

A person's state changes from period to period. To illustrate, given our period length of two years, if a person stays on active duty for another period, active years increase by two, reserve years remain zero, total years increase by two, and the pay grade increases by one (or possibly two, depending on the promotion probabilities) if the person is promoted. If the person leaves active duty to become a reservist, active years do not change, reserve years increase by two, total years increase by two, and the pay grade increases if the individual is promoted. Similarly, if a person is a civilian and remains a civilian, active and reserve years do not change, but total years increase by two. These changes in state are important because they may affect future military, reserve, and civilian pay and future active and reserve retirement benefits.

The active/reserve dynamic retention model is a finite-state, finite-period model. To close it, we must specify what happens at the end of T, the 40th year of work life, age 60. We assume that the individual can no longer serve in the military, and the only "choice" is to be a civilian. We assume that the individual no longer works in the civilian economy after T but receives the expected present value of any military retirement benefits owed him or her over the remainder of his or her life, where the expectation allows for the probability of survival from period to period. The assumption that the individual no longer works at a civilian job could be replaced with an assumption of working, say, until age 67, in which case, the present value of civilian earnings at age 60 would be added to the present value of retirement benefits at age 60. If so, the amount of earnings added would be the same regardless of prior active and reserve choices and would not affect those decisions. Thus, little is lost in omitting civilian earning after age 60.

In summary, the active/reserve dynamic retention model traces an individual from the start of service in an active component until separation from that component. At that point, the individual chooses to hold a civilian job or to join the selected reserve (and hold a civilian job). Those who choose to join the reserves may move back and forth in future periods between reservist and civilian status; those who do not join the reserves must remain civilians in all future periods. In each period, the value of an alternative depends on current earnings, individual preference, the expected value of being able to make the best choice in the next period, and a random shock. Earnings, including possible retirement benefits, are related to the individual's state, which is defined in terms of active years, reserve years, total years, and pay grade. The state changes from period to period, depending on the choices made. The preferences for active duty and reserve duty are constant over time but vary from person to person: Individuals are heterogeneous.

Nested Logit Specification

The active/reserve dynamic retention model describes individual behavior as a series of choices regarding whether to be active, reserve, or civilian, and, having left the active component and initially chosen to be a reservist, to be reserve or civilian. We need to relate the model to data on active and reserve retention and develop an estimation approach. The objective is to obtain an expression from the model of the likelihood of the individual's active or reserve choices over his or her career.

An individual's preferences are assumed to be constant, but the individual's state changes from period to period, so the sequence of past states could matter. However, as mentioned previously, the model is defined as a Markov process. The current state, defined in terms of active years, reserve years, total years, pay grade, and shocks, captures all of the relevant past information, so current choice probabilities do not need to be conditioned on past outcomes. Further, the shocks are assumed to be independent draws that are uncorrelated from period to period. As a result, the probability of an individual's military career—the exact number of years in an active component and the exact sequence of participating in the reserves and being a civilian— can be written as the product of the probabilities of these choices in each period over the work life.

One of the unusual features of our model is the assumption that a reservist holds a civilian job. This is a simplifying assumption, since some reservists are full-time students and some may be unemployed or out of the labor force, but the key idea behind the assumption is that participation in the reserves is often concurrent with another main activity, which we call a *job*. Therefore, a civilian job shock is likely to be present not only in the individual's civilian alternative but also in his or her reserve alternative. To our knowledge, previous applications of dynamic programming to career choices have assumed that shocks are independent across alternatives. This is apart from person-specific fixed effects, such as the active and reserve preferences in our model.

To allow for error correlation between the reserve and civilian alternatives, we modify the model to a nested logit form for the reserve or civilian choice, where the active alternative is one "nest" and the reserve and civilian alternatives represent the other nest. The choice is between the active alternative and the better alternative from the reserve/civilian nest, i.e., the maximum of the reserve alternative and the civilian alternative. To shorten the notation, we rewrite Equation 1 as $Y_{kj}(s_t, \gamma) = V_j + \varepsilon_j$, where V_j represents the nonstochastic terms on the right side, and the state and preference attributes and time subscript have been omitted for brevity. Adapting Ben-Akiva and Lerman's (1985) treatment of the nested logit, we now develop the nested logit specification of the model from the following expressions:

$$V_a + \varepsilon_a$$
$$\max\left[V_r + w_r, V_c + w_c\right] + v_{rc}. \qquad (3)$$

The nested logit model assumes that ε_a has the same distribution as the sum of the errors in the second expression, so we need to ensure that this requirement is met. Also, we assume that all errors are generated from extreme value distributions. When the errors have the same extreme value distribution, and, in particular, when they have the same variance, then the choice between the nests can be shown to have the usual logit form. Train (2003, Chapter 3) provides a proof that, when alternatives have identically distributed, independent extreme-value errors, the probability that a particular alternative is the maximum has the logit form. Ben-Akiva and Lerman (1985) show that the nested logit model can be written as a choice between alternatives, each of which is the maximum choice from its nest. As we show for our model, the errors of these maximum choices can be constructed to have the same variance; hence, Train's proof applies.

The extreme value distribution $EV[a,b]$ has the form $e^{-e^{-(a-x)/b}}$, with mean $a + b\gamma$ and variance $\pi^2 b^2 / 6$, where γ is Euler's gamma (≈ 0.577), a is the location parameter, and b is the scale parameter. The variance is proportional to the square of the scale parameter, and we use the fact that equal scale parameters imply equal variances. Let w_r and w_c be within-nest errors drawn from an extreme-value distribution, $EV[0,\lambda]$, and let v_{rc} be the nest-specific error for the reserve/civilian nest, distributed $EV[0,\tau]$. In other words, v_{rc} can be thought of as a shock that affects both the reserve and the civilian alternatives, whereas w_r and w_c affect each alternative separately.

It is known that $\max[V_r + w_r, V_c + w_c]$ also follows an extreme-value distribution—namely,

$$EV\left[\lambda \ln\left(e^{V_r/\lambda} + e^{V_c/\lambda}\right), \lambda\right].$$

We rewrite the second expression in Equation 3 as follows:

$$\lambda \ln\left(e^{V_r/\lambda} + e^{V_c/\lambda}\right) + w'_{rc} + \upsilon_{rc}, \text{ where}$$

$$w'_{rc} = \max[V_r + w_r, V_c + w_c] - \lambda \ln\left(e^{V_r/\lambda} + e^{V_c/\lambda}\right)$$

$$w'_{rc} \sim EV[0, \lambda]. \tag{4}$$

Define $\varepsilon_{rc} = w'_{rc} + \upsilon_{rc}$. This is the sum of two independent, differently distributed extreme-value variables. The error w'_{rc}. is the single error associated with taking the maximum of $V_r + w_r$ and $V_c + w_c$, and υ_{rc} is the single error at the nest level. The distributions of w'_{rc} and υ_{rc} have the same location parameters (zero) but different scale parameters. In general, the variance of the sum of two independent random variables is the sum of the variances, so the variance of $\varepsilon_{rc} = w'_{rc} + \upsilon_{rc}$ is $\pi^2\left(\lambda^2 + \tau^2\right)/6$, implying a scale parameter of

$$\sqrt{\lambda^2 + \tau^2}.$$

It follows that

$$\varepsilon_{rc} \sim EV\left[0, \sqrt{\lambda^2 + \tau^2}\right].$$

Because we also want ε_a to have the same distribution (i.e., the same location and scale parameters), we assume

$$\varepsilon_a \sim EV\left[0, \sqrt{\lambda^2 + \tau^2}\right].$$

For brevity, let

$$\kappa = \sqrt{\lambda^2 + \tau^2}.$$

Drawing this together, the model may be written as follows:

$$V_a + \varepsilon_a$$

$$\lambda \ln\left(e^{V_r/\lambda} + e^{V_c/\lambda}\right) + \varepsilon_{rc}$$

$$\varepsilon_a, \varepsilon_{rc} \sim EV[0, \kappa]. \tag{5}$$

Assuming that the individual chooses the higher-valued alternative, this leads to a probability for choosing *active* in the usual logit form, as Train (2003) showed:

$$\Pr(active) = \frac{e^{V_a/\kappa}}{e^{V_a/\kappa} + e^{\lambda \ln\left(e^{V_r/\lambda} + e^{V_c/\lambda}\right)/\kappa}}$$

$$= \frac{e^{V_a/\kappa}}{e^{V_a/\kappa} + \left(e^{V_r/\lambda} + e^{V_c/\lambda}\right)^{\lambda/\kappa}}. \tag{6}$$

The second line follows from the fact that $e^{b \ln a} = a^b$.

The within-nest error terms, ω, are distributed $EV[0, \lambda]$, and the "total" error terms, ε, are distributed

$$EV\left[0, \sqrt{\lambda^2 + \tau^2}\right].$$

Therefore, the within-nest, choice-specific portion of the total error accounts for the following fraction of the error variance:

$$\frac{\lambda^2}{\tau^2 + \lambda^2}. \tag{7}$$

It follows that the fraction of the error variance attributable to the within-nest common shock is one minus this amount, or $\tau^2 / (\tau^2 + \lambda^2)$.

Stated differently, we can think of the problem of selecting the best alternative from the nest as choosing between

$$V_r + \omega_r + \upsilon_{rc}$$
$$V_c + \omega_c + \upsilon_{rc}. \tag{8}$$

It follows that the correlation between these two total utilities is

$$\rho = \frac{Cov\left[V_r + \omega_r + \upsilon_{rc}, V_c + \omega_c + \upsilon_{rc}\right]}{\sqrt{Var\left[V_r + \omega_r + \upsilon_{rc}\right]}\sqrt{Var\left[V_c + \omega_c + \upsilon_{rc}\right]}}$$

$$= \frac{\tau^2}{\tau^2 + \lambda^2}. \tag{9}$$

As shown in Equation 9, a larger variance of the common shock, υ_{rc}, results in a larger correlation between the reserve and civilian alternatives. Thus, the nested logit formulation succeeds in giving us a specification that allows the shocks to the reserve and civilian alternatives to be correlated, and the greater the common shock, the greater the correlation.

Applying the rule for the distribution of the maximum of two values, we see that

$$\max\left[V_a + \varepsilon_a, \lambda \ln\left(e^{V_r/\lambda} + e^{V_c/\lambda}\right) + \varepsilon_{rc}\right]$$
$$\sim EV\left[\kappa \ln\left(e^{V_a/\kappa} + e^{\lambda \ln\left(e^{V_r/\lambda} + e^{V_c/\lambda}\right)/\kappa}\right), \kappa\right]$$
$$EV\left[\kappa \ln\left(e^{V_a/\kappa} + e^{\lambda \ln\left(e^{V_r/\lambda} + e^{V_c/\lambda}\right)/\kappa}\right), \kappa\right]$$
$$= EV\left[\kappa \ln\left(e^{V_a/\kappa} + \left(e^{V_r/\lambda} + e^{V_c/\lambda}\right)^{\lambda/\kappa}\right), \kappa\right]. \tag{10}$$

As before, the last line follows from $e^{b \ln a} = a^b$. Now, using the formula in Equation 10 for the mean of an extreme-value distribution, the expected value of the maximum of the two alternatives, active versus the maximum of reserve/civilian, is

$$\kappa\left(\gamma + \ln\left[e^{V_a/\kappa} + \left(e^{V_r/\lambda} + e^{V_c/\lambda}\right)^{\lambda/\kappa}\right]\right). \tag{11}$$

Further, the expected value of the maximum of the two alternatives, reserve and civilian, given that an individual has left the active component and cannot reenter it, is

$$\kappa\left(\gamma + \ln\left[\left(e^{V_r/\lambda} + e^{V_c/\lambda}\right)^{\lambda/\kappa}\right]\right)$$
$$= \kappa\gamma + \lambda\ln\left[e^{V_r/\lambda} + e^{V_c/\lambda}\right]. \tag{12}$$

The first line of Equation 12 does not contain the term $e^{V_a/\kappa}$ because the constraint that the individual cannot reenter the active component means, in effect, that V_a is set to negative infinity, and $e^{-\infty} = 0$. The second line of Equation 12 simplifies the log expression.

The expected value of the maximum of a set of choices is referred to as the *surplus function*, and the surplus function can be used to derive choice probabilities. The Williams-Daly-Zachary theorem (see McFadden, 1981) states that the probability of choosing a given alternative equals the partial derivative of the surplus function with respect to the value of the alternative. Thus, the probability of choosing to remain in an active component is as follows:

$$\Pr(active) = \frac{\partial \kappa\left(\gamma + \ln\left[e^{V_a/\kappa} + \left(e^{V_r/\lambda} + e^{V_c/\lambda}\right)^{\lambda/\kappa}\right]\right)}{\partial V_a}$$
$$= \frac{e^{V_a/\kappa}}{e^{V_a/\kappa} + \left(e^{V_r/\lambda} + e^{V_c/\lambda}\right)^{\lambda/\kappa}}. \tag{13}$$

This is the same as that shown in Equation 6, which replicated the usual logit specification. To emphasize the meaning of Equation 13, we restate it as

$$\Pr\left(V_a + \varepsilon_a > \lambda \ln\left(e^{V_r/\lambda} + e^{V_c/\lambda}\right) + \varepsilon_{rc}\right) = \frac{e^{V_a/\kappa}}{e^{V_a/\kappa} + \left(e^{V_r/\lambda} + e^{V_c/\lambda}\right)^{\lambda/\kappa}}.$$

By the same approach, the probabilities of choosing reserve or civilian are

$$\Pr(reserve) = \frac{\left(e^{V_r/\lambda} + e^{V_c/\lambda}\right)^{\lambda/\kappa}}{e^{V_a/\kappa} + \left(e^{V_r/\lambda} + e^{V_c/\lambda}\right)^{\lambda/\kappa}} \frac{e^{V_r/\lambda}}{e^{V_r/\lambda} + e^{V_c/\lambda}}$$

$$\Pr(civilian) = \frac{\left(e^{V_r/\lambda} + e^{V_c/\lambda}\right)^{\lambda/\kappa}}{e^{V_a/\kappa} + \left(e^{V_r/\lambda} + e^{V_c/\lambda}\right)^{\lambda/\kappa}} \frac{e^{V_c/\lambda}}{e^{V_r/\lambda} + e^{V_c/\lambda}}. \tag{14}$$

The probabilities of choosing reserve or civilian, given that the individual has left the active component and cannot reenter it, are, respectively,

$$\Pr(reserve \mid not\ active) = \frac{\partial\left(\kappa\gamma + \lambda \ln\left[e^{V_r/\lambda} + e^{V_c/\lambda}\right]\right)}{\partial V_r} = \frac{e^{V_r/\lambda}}{e^{V_r/\lambda} + e^{V_c/\lambda}}$$

$$\Pr(civilian \mid not\ active) = \frac{\partial\left(\kappa\gamma + \lambda \ln\left[e^{V_r/\lambda} + e^{V_c/\lambda}\right]\right)}{\partial V_c} = \frac{e^{V_c/\lambda}}{e^{V_r/\lambda} + e^{V_c/\lambda}}. \tag{15}$$

A comparison of Equations 14 and 15 shows that the probability of choosing to be a reservist equals the probability of choosing the reserve/civilian nest multiplied by the probability of choosing reserve, given that the individual is in the nest (i.e., has left active duty and cannot reenter). A similar statement holds for the probability of choosing the civilian outcome.

In summary, we obtained expressions for the value of the value function for each alternative in a given period. These expressions divided the value into a nonstochastic part that consisted of current pay, the monetary value of the preference for the activity, and the expected value of being able to choose the highest-valued alternative in the next period, as well as a random shock. We argued that the random shocks for the reserve and civilian alternatives might be correlated, and we allowed for this by adding a common shock to these alternatives. Assuming that the shocks were drawn from extreme-value distributions, as described earlier, we presented expressions for the surplus functions of the active versus reserve/civilian choices and the reserve versus civilian choices as conditional on having left the active component. The surplus functions represent the expected utility of being able to choose the maximum. We applied the Williams-Daly-Zachary theorem to obtain expressions for the probability of remaining on active duty and for the probabilities of choosing reserve or civilian, respectively, conditional on having left the active component. Because the model sets up the choice process as a Markov model in which past history is fully summarized in the current state, and because the shocks are independent from period to period, the choice probabilities in each period contain all relevant information for the individual's choice and are independent from period to period. Consequently, the choice probabilities can be expressed for each choice that an individual makes in each period and multiplied together to obtain an expression of the likelihood of observing the individual's exact choice sequence. This provides a pathway from the theoretical model to its empirical application.

The model assumes that individuals differ in their preferences for active or reserve service. The likelihood of a given individual's sequence of choices is conditional on the individual's preferences. However, one of the challenges in estimating the model is that preferences are not directly observed. We address this challenge by estimating the parameters that represent the distribution of tastes across individuals such that the individual-level choice sequences generated by the distribution are consistent with those observed in the data.

A second challenge concerns estimating the expected value of the maximum. The expected value of the maximum in a given period depends on the expected value of the maximum in the next period, and the one after that, and so on. Thus, the expected value of the maximum embeds a recursive sequence of optimizing decisions in all future periods, ending at the final period of work life, T. This is a logically consistent way of allowing future pays, retirement benefits, and uncertainty (unrealized shocks) to enter current decisionmaking: In other words, it is a way of modeling that allows the individual to look ahead and incorporate information known currently about future pay, promotion probabilities, benefits, civilian wages, and uncertainty, and it assumes that the individual will not be bound by today's choice but will reoptimize in each future period through T, depending on the circumstances realized in those periods.

Appendix B describes the methods used to estimate the model and the data.

Simulation

Once the model has been estimated, it can be used for policy simulations. The first step in conducting a simulation is to create a population of synthetic individuals. Within the context of the model, an individual is an entity with specific preferences for active and reserve service and a specific set of random shocks for each alternative in each period. Therefore, the simulation creates the individual by a random draw of active and reserve preferences from the preference distributions and a set of random draws from the shock distributions. Consistent with the model, the individual is assumed to know his or her preferences, the values of the shocks in the current decision period, and the distributions of the shocks (i.e., the scale parameters τ and λ, which are used in the individual's computation of the expected value of the maximum in the next period). That is, even though the analyst knows the shocks for each period in the individual's work life, in any period, the individual does not know the values of shocks in future periods.

The second step in a simulation is to specify the compensation structure. Our policy analysis generally focuses on comparisons between the current compensation structure and alternative structures. The current structure was coded into the model when it was estimated; the observed active and reserve retention behavior was conditional on the current compensation structure. New coding is required for each alternative structure. The policy alternatives under consideration involve changes in vesting, period of first receipt, and amount of retirement benefits, as well as the provision of gate pay (payable upon completion of a certain number of years of service), separation pay, and U.S. Department of Defense (DoD)–paid DCs.

The third step is to put the synthetic population into the model; compute each person's value functions recursively, as described previously; and let the person choose an alternative at each decision point. The result is a career path that is optimal for the individual, given the compensation structure and particular shocks he or she faced in each period.

The fourth and final step is postprocessing. The career information for our synthetic individuals includes period-to-period data on their state (active years, reserve years, total years, and pay grade). We combine this to create information about the synthetic population (its active-duty retention, participation in the selected reserve, highest grade attained, expected years of active and reserve service, and compensation cost). We can manipulate this information to make cost comparisons subject to holding active-duty personnel strength constant, or, alternatively, to make strength comparisons subject to holding cost constant. When discounting is required, we use the discount factor appropriate for the calculation, i.e., the individual's discount factor for calculations from the perspective of the individual or the organizational discount factor for calculations from the organization's perspective.

The cost concept that we use is current cost, rather than life-cycle cost. Life-cycle cost is often used in weapon system procurement costing, and it could be used in manpower costing if "procuring" a cohort of new entrants is considered purchasing a new asset. However, policymakers are accustomed to viewing manpower costs as current outlays, so current costing seems more appropriate. For an earlier version of this monograph, we computed the life-cycle costs as well. Computing life-cycle costs involves discounting future compensation to compute the present discounted value. Our results are qualitatively similar, but life-cycle costs are lower than current costs.

With respect to our simulation, we simulate the career behavior of a population entering active duty, and, to convert our results into a current setting, we adopt the assumption that the personnel force is in a steady state. By implication, the active and reserve retention behavior that we simulate can be interpreted as the force structure that one would see in the cross-section, i.e., in the current period. Under this assumption, we compute two costs: current compensation and deferred compensation. Current compensation includes regular military compensation plus any gate pays and separation pays. Deferred compensation includes outlays required to fund DBs and DCs for vested personnel upon their departure from service. For example, if a service member left active duty after completing 24 years of service, we register a cost equal to the present discounted value (PDV) of the stream of retirement benefits expected to be paid to the individual in all future periods, under the terms of the retirement-benefit system we were simulating, and allowing for survival probabilities.

Limitations

The model provides a great deal of richness and realism in its ability to capture complex decisions that depend on intertemporal comparisons in a world with uncertainty while allowing the individual the flexibility to change course in the future. The model is valuable in that it includes both active and reserve decisions and that its application to evaluating compensation reforms is empirically grounded. But the model has limitations, and, before we proceed to the results, we want to discuss some of these.

The model begins with service members who have entered an active component. The model does not include recruiting or the cost of recruiting and training. However, when we simulate alternative compensation proposals, the result could be a force with longer or shorter careers than the current force, and this can affect the quantity of recruits needed to sustain a force of a given size or cost. For example, if careers were longer but the force size was constant, fewer recruits would be needed, and this would reduce recruiting and training cost. In contrast, if careers were shorter, more recruits would be needed, and these costs would increase. Similarly, if a service branch lengthened some careers and shortened others while holding force size constant, there might be no effect on recruiting cost.

Because the model focuses on service members who enter active duty, it does not address service members who enter the reserves directly (without prior service in an active component). However, the model can be modified to handle a "pure reserve career" by eliminating the active career. Non–prior service reservists can be followed from the beginning of their reserve service though the remainder of their work lives. As with the current model, one might want to extend the pure reserve model to include enlistment decisions.

Both the formulation of the model as described in this chapter and the method that we use to estimate it assume time stationarity in military and civilian pay. In the 1990s, military pay declined relative to civilian pay until the end of the decade, when the National Defense Authorization Act for Fiscal Year 2000 (Public Law 106-65) mandated a significant increase in military pay. Because the model currently assumes pay stationarity, it will not fit the data as well as a model allowing for pay change. However, retention remained high in most specialties during the 1990s, despite the decline in military pay, though there were pockets of retention decrease in high-tech skills. The main effect of the decline in relative pay occurred in recruiting in 1998 and 1999, and as mentioned, the model does not cover recruiting. The fairly steady level of retention during the 1990s and into the 2000s suggests that our pay-stationarity assumption reflects historical trends. Reenlistment bonuses helped to sustain retention in the 1990s and 2000s. Reenlistment bonuses and other S&I pays are not included in our pay measure, but their effect enters the model indirectly through the shock terms. The payment of a bonus would contribute to a positive shock.

The model does not include demographic characteristics, such as age, education, race, ethnicity, gender, and marital status. These can be included in future work. For instance, age at entry, race, ethnicity, and gender can be modeled as observables in what are now the mean preferences for active duty and reserve duty. Marital status might be modeled as a Markov process embedded in the value functions. The effect of a change in the structure of military compensation might affect single service members differently than married service members, and an expanded model would allow one to test for this, as well as for the effects of the other demographic variables.

The model assumes age-independent utility. The utility function in the model depends on current compensation, the implicit monetary value of the preference for active service or reserve service, the implicit monetary value of the shock, and the discounted value of the expected value of making the best choice in the next period. However, it is possible that utility varies with age, and the inclusion of age could differentially affect the transition probability from one state to the next under the current compensation system versus the QRMC proposal. While this possibility deserves future exploration, our sense is that differential age effects under current versus QRMC compensation systems are likely to be small because the proposals are fairly close in value to one another. Therefore, although we do not have age-dependent utility, we think the simulated differences in retention, man-years, and cost between the current and QRMC compensation systems are likely to be similar to those that would be achieved under an age-dependent utility specification.

The model assumes that promotion probabilities are invariant to policy changes. This is plausible given the compensation policies that we consider, because they do not result in radical changes in retention. DoD and service policy on promotions is to make promotion timing and promotion probability predictable so that service members can plan their careers and anticipate their advancement through the ranks. The model, however, does allow for high-year-of-tenure rules (see Appendixes C and D). Under the current and alternative compensation schemes, these rules prevent the emergence of too many service members in the highest ranks.

The simulations that we present, which are based on the estimated parameters, compare two steady states: the active and reserve forces under the current compensation policy and under an alternative compensation policy. We do not simulate the transition between the steady states. The transition is a study in itself, because it will depend on transition policy. For instance, will a new policy be phased in gradually by applying it only to recruits who enter after a certain date, or will it cover all personnel immediately? Will personnel have a choice between the current policy and the new policy, and if so, what will the choice entail?

The model does not incorporate deployment, deployment-related pay, or reserve activation and deployment. In particular, if deployment during the operations in Iraq and Afghanistan had reduced retention, then the absence of deployment in the model could result in biased parameters. Military pay might appear to have a smaller effect on retention than it actually had, because (by assumption) deployment was reducing retention. But the fact is that retention has remained quite steady during these operations, and bonuses and deployment-related pays have contributed to sustaining it. For this reason, we do not think that our estimates have been seriously affected by the absence of deployment in the model, though we do think that the model should be extended to include deployment.

We have described these limitations because they are facts that should be known about the model. In considering the policy implications of the model, it is therefore crucial to keep in mind that, while the model has important strengths (e.g., it assumes that people are forward-looking and rational), it simplifies some details of military compensation and military life. It seems plausible, therefore, that the policy implications that we draw are not very sensitive to these assumptions, but the alternative—namely, that some of the policy implications are quite sensitive—is also possible. Many if not all of them can be addressed in future work as needed, so we defer further exploration to follow-on analyses. Nonetheless, in light of the model's limitations, it may also be worth pointing out that no other model has the capability of the active/reserve dynamic retention model. One-period, single-equation models of retention at first term or second term entirely miss retention behavior later in the career. The annualized cost-of-leaving (ACOL) model, to its credit, has been extended to allow for individual heterogeneity and does consider deferred compensation, but as Gotz (1990) notes, it is not intertemporally consistent, and it is inadequate in its handling of future uncertainty. Neither the single-equation nor the ACOL model follows personnel from the actives to the reserves, and, thus, neither is capable of showing the full effects of compensation on the active and reserve forces. The active/reserve dynamic retention model, in contrast, presents a unified, intertemporally consistent framework for analyzing the effect of compensation on active and reserve personnel, and it allows for heterogeneity of preferences with respect to both active duty and reserve duty. Finally, the model presents a novel, but appropriate, treatment of the reserve and civilian choices by placing them within nests. This recognizes that an individual who serves in the reserves is affected by many of the same factors as a civilian. Therefore, although the model has limitations and room for improvement, it comprises a cohesive and effective framework for compensation analysis.

CHAPTER FOUR

Estimates and Base-Case Results

This chapter presents the parameter estimates and simulation results for the base case, which is the current compensation system. We compare the base-case results with statistics obtained from our data set that act as benchmarks to assess the fit of our model.

Parameter Estimates

Table 4.1 presents estimates of the active/reserve dynamic retention model. The estimates are denominated in thousands of dollars. The reference population for the estimates consists of individuals who have just begun their first term of service in the active-duty enlisted force. The standard errors of the estimates of the means, the alpha parameters, and the scale parameters are based on the method of Berndt et al. (1974). The standard error of the square root of the sum of alpha-21 squared and alpha-22 squared is based on the delta method. The delta method provides standard-error estimates of a nonlinear transformation of the parameter estimates, where the transformation is derived from the Cholesky equations (see Appendix B). The method involves expanding the transformation by a Taylor approximation and taking the variance, where the variance of the transformation is a function of the variance of the parameter estimates (see, for example, Fieveson, 2005.)

Table 4.1
Parameter Estimates for Enlisted Personnel, by Service
(standard errors in parentheses)

Parameter	Army	Navy	Air Force	Marine Corps
Active taste mean μ_a	−12.209 (0.08)	−15.598 (0.07)	−10.263 (0.06)	−9.942 (0.10)
Reserve taste mean μ_r	−31.442 (0.09)	−34.479 (0.10)	−28.432 (0.07)	−30.881 (0.11)
Alpha-11	0.051 (2.13)	0.882 (0.72)	0.731 (1.04)	0.551 (3.65)
Alpha-21	1.329 (1.73)	1.128 (3.07)	1.400 (1.38)	1.848 (1.34)
Alpha-22	2.479 (0.90)	2.320 (0.88)	0.172 (8.01)	0.0337 (61.89)
Square root of (alpha-21 squared + alpha-22 squared)	2.813 (1.86)	2.580 (3.23)	1.411 (5.14)	1.849 (13.65)
Scale parameter τ	6.814 (4.18)	11.342 (2.85)	6.932 (2.81)	9.128 (2.84)
Scale parameter λ	38.246 (0.11)	39.827 (0.12)	32.319 (0.09)	34.677 (0.13)

The parameter estimates of the mean active and mean reserve preferences are fairly similar across the services. The means of the active and reserve preferences are highly statistically significant, judging by the fact that they are far more than twice the standard error. A negative average preference among individuals who chose to enter active duty reflects the relatively short active-duty careers of many who join and provides a way of accounting for the relatively early exit of a large segment of personnel, in addition to the other factors in the model. The average preference for reserve service is lower than the average preference for active duty and reflects a lower propensity to join the reserves than to join the actives, which is not surprising, given that many who leave active duty do not join the reserves. The lower reserve preference also reflects a greater tendency to exit from the reserves than from the actives.

The alpha parameters are related to the standard deviation of the active and reserve preferences and the correlation between the preferences, as discussed in Appendix B. The alpha-11 parameter is the standard deviation of the preference for active duty. The standard deviation is small relative to the mean active-duty preference for each of the services, indicating that preferences are fairly tightly distributed around the mean. The standard deviation of the preference for reserve duty equals the square root of the sum of alpha-21 squared and alpha-22 squared, so we include a separate entry for this quantity. It is small relative to the mean preference for reserve service. The findings imply that individual heterogeneity in preferences for active and reserve service plays a relatively small role in retention. The small values of the standard deviations of the preferences relative to the means in these estimates may be a result of limited variation in individual-level data. The model does not include demographic variables, and we do not have data on person-specific wage opportunities in the civilian sector. For example, service members with persistently higher civilian wage opportunities would tend to have a lower preference for the military, but the estimates cannot address this because we have used the same civilian wage by age for all individuals.

The scale parameters provide information on the standard deviations of the common random shock for the reserve/civilian nest, as well as within–civilian/reserve nest shocks. Further, these two scale parameters provide information on the standard deviation for the choice from each nest—active and reserve/civilian—as discussed in Chapter Three, Equation 7. The random shocks account for transitory, unobserved factors that influence the decisions in the model. A person with a high preference for active duty could nevertheless choose to leave active duty when faced with a large negative shock. Similarly, a person with a relatively high (or not-so-negative) preference for reserve service is more likely to join the reserves when a positive random shock occurs. The standard deviations of the shocks are all highly statistically significant. Also, based on inspection, we see that the tau shock and lambda shock scale parameters have considerably different values and are no doubt statistically different from one another. The difference between these estimates supports our use of the nested logit specification in the dynamic retention model.

Table 4.2 draws together information on the standard deviation of preferences, the correlation between preferences, the standard deviations of the within-nest common shock, the shock specific to each choice within the nest, the overall shock for the nest, and the correlation of the shocks between the reserve and civilian choices (see Equation 9 in Chapter Three).

As mentioned, the standard deviation of active-duty preference is small—less than $1,000—for all the services. This is less than one-tenth of the mean active preference (shown in Table 4.1). The standard deviation of reserve preference is several times larger, but it is only about one-tenth the mean reserve preference. The correlation between active preference and reserve preference ranges from about 0.44 to 1.00; in other words, there is a high correlation between preferences. The wide range of this correlation across the services is consistent with the lack of statistical precision of the estimate, i.e., we found large standard errors. Although the absence of large and statistically significant standard deviations for the preferences may seem like a non-result, it underscores the importance of actually estimating the model

Table 4.2
Standard Deviations of Active Preference, Reserve Preference, Correlation Between Active and Reserve Preferences, and Nest Shocks

Standard Deviation or Correlation	Army	Navy	Air Force	Marine Corps
Active preference	0.051	0.882	0.731	0.559
Reserve preference	2.813	2.580	1.411	1.849
Correlation between preferences	0.472	0.437	0.993	0.999
Within-nest common shock	8.739	14.547	8.890	11.707
Within-nest–specific shock	49.05	51.08	41.45	44.56
Nest shock	49.82	53.11	42.39	48.17
Correlation between reserve and civilian shocks	0.031	0.075	0.044	0.144

NOTE: Standard deviations are denominated in thousands of dollars.

rather than presuming that the standard deviations would be large, based on an assumption that there are large differences in taste for military service across service members. As mentioned, it is possible that future estimation of the model will detect heterogeneity, depending on the data available.

With respect to the shocks, the within-nest-specific shock accounts for most of the variance in the overall nest shock. The correlation between the shocks for the reserve and civilian choices is low, ranging from 0.03 to 0.14. (In general, the scale for the correlation is –1 to 1.) This means that the shocks affecting the civilian job (or education) and reserve service are largely, but not entirely, independent.

Base-Case Results

Figure 4.1 is based on simulation results in the base case for the Army, given the parameter estimates in Table 4.1. The figure shows the number of active members who are present at each year of service, where years are counted as years since starting active duty, for a cohort of 5,000 simulated individuals entering active duty at time zero, scaled

Figure 4.1
Active-Duty Enlisted Members, by Year of Service:
Army Base-Case Simulation

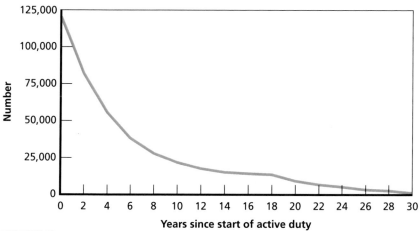

RAND MG764-4.1

up to an enlisted strength of 415,000. Under the base case, the retirement system is assumed to be the high-three system. In the simulation, 10.5 percent survive until 20 years of service. Expected active-duty man-years per accession in the base case is 7.0 years. The retention profile implies an active-duty accession requirement of 118,966 over two years—recall that decisions are made every two years in our model—or 59,483 per year in the initial period.

The simulation for the Army base case assumes a personal discount rate of 15 percent, below the average personal discount rate estimated by Warner and Pleeter (2001). This rate was chosen because it fit the data better than alternative assumptions in terms of the force profile and the percentage of personnel who reach 20 years of active service. To illustrate, Figure 4.2 shows the simulated Army enlisted active-duty personnel profile, assuming a force size of 415,000, under alternative personal discount rate assumptions. Imposing a rate of 12.5 percent or a rate of 10 percent in the estimate yields forces that produce greater retention and fewer accessions. The fraction of the entering cohort that becomes vested at YOS 20 is 17.8 percent, assuming a

Figure 4.2
Active-Duty Enlisted Members, by Year of Service:
Army Base-Case Simulation with Alternative Personal Discount Rate
Assumptions

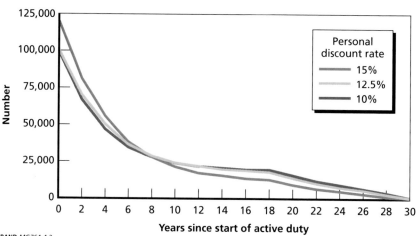

personal discount rate of 12.5 percent, and 19.3 percent, assuming a rate of 10 percent. These vesting rates are high for the Army. The 10.4-percent vesting rate associated with a personal discount rate of 15 percent seems more reasonable.

Figure 4.3 shows the simulated active-duty YOS profile for the base case for each service, given our parameter estimates and assuming

Figure 4.3
Active-Duty Enlisted Members, by Year of Service:
Base-Case Simulation for All Services

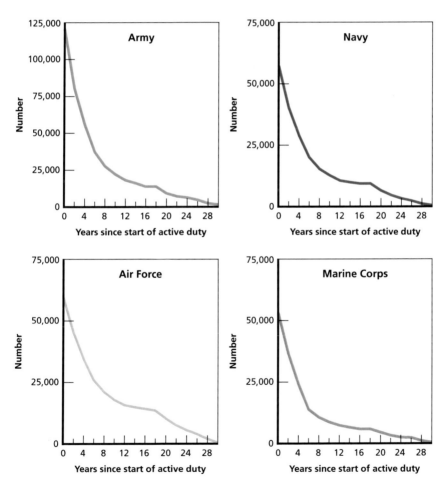

an active enlisted strength of 415,000 for the Army, 280,000 for the Air Force, 180,000 for the Marine Corps, and 225,000 for the Navy. Table 4.3 presents relevant statistics with respect to active-duty outcomes for each service's base case. We find that 14.8 percent of Navy accessions reach YOS 20, while 22.4 percent of Air Force accessions reach YOS 20. For the Marine Corps, the percent vesting for active retirement is 10.0 percent.

Figure 4.4 shows the reserve-component simulation results for the base case for each service, e.g., the sum of the Army Reserve and Army National Guard, in the case of the Army. The figure is based on the same cohort of 5,000 simulated individuals and tracks their participation in the selected reserve after leaving active duty. Years are counted as years since starting active duty, and the first individuals enter the reserves after serving two years in the actives. In the case of the Army, for example, of the 5,000 active members, 19.1 percent join the reserve components at some point during the 40-year career path. Conditional on joining the reserves, 12.6 percent vest in the reserve retirement system. That is, they achieve 20 years of creditable service, with the last six served in the reserve component. The number in the reserves declines in the last 15 years of the 40-year work life, likely indicating that more and more reservists vest during this phase.

Table 4.3
Base-Case Active-Duty Outcomes

Service	Percent Who Qualify for Active Retirement	Average Active Years of Service per Accession	Assumed Enlisted Strength	Active Accessions Required to Maintain Strength
Army	10.5	7.0	415,000	59,483
Air Force	22.4	9.5	280,000	29,380
Navy	14.8	7.7	225,000	29,118
Marine Corps	10.0	6.7	180,000	26,059

NOTE: Simulations based on 5,000 cases.

Figure 4.4
Enlisted Members in the Selected Reserve, by Year of Work Life: Base-Case Simulation of Personnel Who Began in an Active Component, All Services

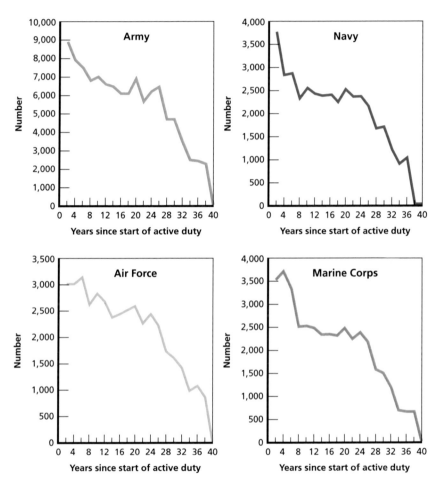

Table 4.4 provides cost results from the simulation. The costs shown in the table are based on the active enlisted end strengths shown in Table 4.3. Given the active-duty retention behavior of the simulated individuals, total active-duty cost per man-year for the Army is $46,346. With a force size of 415,000, total active-duty costs are $19.2 billion.

Table 4.4
Base-Case Active-Duty Cost Outcomes

Service	Expected Active Years of Service per Accession	Average Active-Duty Cost per Man-Year ($)	Assumed Strength	Total Active-Duty Cost ($ billions)
Army	7.0	46,346	415,000	19.2
Navy	7.7	49,194	225,000	11.1
Air Force	9.5	34,639	280,000	9.7
Marine Corps	6.7	46,780	180,000	8.4

The costs are calculated as described in the Chapter Three and include current military compensation—specifically, regular military compensation, including the federal tax advantage, an estimate of active-duty retention bonuses, and an estimate of reserve-affiliation bonuses, plus the amount needed to fund retirement benefits. The latter is estimated by discounting expected future retirement benefits for those who qualify for them at a real government discount rate of 3.5 percent.

Table 4.5 shows summary statistics with respect to the reserves, as well as reserve cost results. The table shows that the percent of active-duty personnel who join the reserves varies across services.[1] It also shows the number of reserve accessions, among those with prior active service, to sustain reserve strength, given the assumed size of the active component, and retention patterns in the reserve components among those with prior active service. Reserve cost per man-year, among those with prior active service, is on the order of about $7,000, though somewhat lower for the Navy. Given the total strength for each service (among those with prior active service), total reserve costs are higher for the Army, due to its larger size, than for the other reserve components.

---

[1] Note that the prior-service accession rates in Table 4.5 include only those with prior active service who join the reserves with no prior reserve service. These figures are not directly comparable to standard published prior-service accession rates, because they do not include the accession of those with prior reserve service who do not also have active service. Since a sizable fraction of prior-service reserve accessions are those with prior reserve service, the figures in Table 4.5 appear relatively low compared to prior-service accession rates that include reserve accessions among those with and without active service.

Table 4.5
Base-Case Reserve-Duty Cost Outcomes

Service	% Who Ever Join the Reserves	Expected Reserve Years of Service per Accession	Expected Reserve Prior Active-Service Accessions Required	Average Reserve-Duty Cost per Man-Year ($)	Assumed Strength	Total Reserve-Duty Cost ($ billions)
Army	21.62	8.31	12,860	6,972	106,879	0.745
Navy	17.48	7.88	4,979	6,298	39,239	0.247
Air Force	18.10	7.75	5,318	7,047	41,214	0.290
Marine Corps	19.96	7.50	5,352	5,801	40,123	0.233

Benchmarking of Base-Case Results

One way to assess the usefulness of the model in simulating the effects of policy outcomes is to consider how well the model's predictions compare with the actual data. While we do not conduct formal goodness-of-fit tests, such comparisons provide information on how well the model captures reality. We informally assess model fit by comparing results from the base-case simulations to computations based on the WEX analysis file. Specifically, we compare the percent of active-duty leavers who become vested, the percent of members with prior active service who join the reserves, and the percent of such reserve accessions who qualify for reserve retirement. These percentages are used as benchmarks to assess model fit.

Because our WEX analysis file spans 1990 to 2007, or 18 years, we must approximate the percent who vest under the active-duty retirement system in our data as the fraction of entrants who reach 18 rather than 20 years of service. The percent who reach 18 years will be slightly higher than the percent who would reach 20 years, because a few members might separate between YOS 18 and 20. Thus, the estimate from the data will slightly overstate the vesting rate. In the case of the percent of active-duty leavers who ever join the reserves, to make the computations using the simulation and the WEX data more comparable, we compute the percent who ever join the reserves among those who separate from active duty by YOS 18, rather than YOS 30. In the case of the percent of reserve accessions who qualify for reserve retirement, the 1990–2007 WEX data will understate this percent, because some individuals might qualify for reserve retirement beyond the 18 years of our data, and we do not capture these retirements. To address the issue of insufficient time within 18 years to qualify for reserve retirement, we extended the WEX analysis file, covering the period 1990 to 2007, to include observations of individuals who entered active duty prior to 1990 and who thus could have up to 40 years of active and reserve service. The WEX data provided by the Defense Manpower Data Center include these entrants, though they were not used in the estimation analysis file because their retention behavior is conditional on staying until 1990, by definition of how the WEX file is constructed, and their inclusion in estimation could bias our estimates. Thus, the data for the computation of this benchmark include entry cohorts that started active duty prior to 1990 and stayed until 1990, either in the active or reserve components. The simulations span an entire 40-year active and reserve career. Thus, extending the data through which we compare model fit to include longer careers makes our data more comparable with the simulation concept. We used these data to compute the fraction of individuals transitioning in each period from reserve to civilian, civilian to reserve, reserve to reserve, and civilian to civilian in periods after active duty and used the transition probabilities to obtain a benchmark for the fraction of reservists qualifying for reserve retirement benefits.[2]

As noted in Chapter Three, although the data do not cover a full 20 years, we are able to estimate parameters of the distributions of taste for service and the random shocks that give rise to the observed retention patterns for the 18 years of data that we do have. Thus, the data can be used to estimate the model parameters that underlie the retention decisionmaking process and span an entire career. These estimated

---

[2]   In work not reported here, we attempted to include the 1990 observations when estimating our model, but the resulting parameter estimates were implausible. Our method did not adequately control for selectivity of individuals already in the military as of 1990. As a result, we estimated the model strictly with data on new entrants into active duty. Even though our selectivity adjustment was unsuccessful, the 1990 data were nevertheless valid for computing empirical transition probabilities.

parameters are then used as inputs in the simulation of the base case and policy alternatives. The comparison of predictions of the model with the percentages in the data gives an indication of model fit.

Table 4.6 presents the benchmarks, obtained from the WEX analysis file, with the percentages obtained from the base-case simulations, shown in Tables 4.4 and 4.5.

The model appears to fit the data reasonably well in terms of the active-duty benchmark, specifically, the percentage reaching 20 years of active service. The retention profiles predicted by the models result in the percentage of active-duty service members who qualify for active-duty retirement benefits as follows: Army, 10.5 percent; Navy, 14.8 percent; Marines, 10.0 percent; and Air Force, 22.4 percent. In the WEX data, these percentages are 10.9, 17.1, 9.2, and 26.3, respectively.

The model somewhat underestimates the percentage who ever join the reserves following active-duty separation. For example, the model predicts that 23.5 percent of Army active-duty leavers (by YOS 18) ever join the reserves, while the analysis of the WEX data yields an accession rate of 29.7 percent. Thus, the fit could be improved in terms of the reserve accession rate. The same is true for the reserve retirement rate among those who join the reserves from active duty. The model predicts that 19.0 percent of Army reservists joining from active duty qualify for reserve retirement. For the Army, analysis of the WEX data yields an estimate of 14.5 percent. For the Air Force, the model predicts a retirement rate of 19.3 percent, while the WEX data reveal a rate of 21.9 percent.

Because the fit of the active-duty model seems reasonable, based on our informal comparison, the focus of the policy analysis in Chapter Five is on active-duty outcomes. Furthermore, the results of the analysis provide a basis for future study, such as a pilot test, to the extent that alternative proposals, such as the QRMC proposal, appear promising. Future work will continue to refine the model to improve the fit of the reserve side.

Table 4.6
Base-Case Simulation Results Versus Benchmarks from the WEX Analysis File

Service	Simulation			WEX Data		
	% of Active-Duty Entrants Who Qualify for Active Retirement	% of Active-Duty Leavers Who Ever Join Reserves	% of Reserve Accessions Who Qualify for Reserve Retirement	% of Active-Duty Entrants Who Qualify for Active Retirement	% of Active-Duty Leavers Who Ever Join Reserves	% of Reserve Accessions Who Qualify for Reserve Retirement
Army	10.5	23.5	19.0	10.9	29.7	14.5
Air Force	22.4	22.8	19.3	26.3	29.6	21.9
Navy	14.8	20.1	18.5	17.1	29.4	17.3
Marine Corps	10.0	21.7	12.1	9.2	33.2	13.2

CHAPTER FIVE

Simulation Results

This chapter presents the results of the simulation of the policy alternatives requested by the 10th QRMC. Because the DACMC provided the
foundation for the QRMC proposal, we first simulate the effects of the
DACMC proposals using our model estimates for the Army. We consider the Army because the results for this service illustrate the general
direction of our results for these proposals. We then present simulation
results for the QRMC proposal for all services. The focus of the discussion is on the active-duty results, because the effects on active-duty
personnel are likely to be effects that sell or preclude a given proposal,
though the effects on the reserves are noted in some cases. We are
currently preparing a companion report on reserve retirement reform,
which applies the model more specifically to the reserves. We consider
the results for all services because each service is likely to require information about the effects of the QRMC proposal.

DACMC Results

The DACMC proposals were the starting point for the QRMC delib-
erations on retirement reform. We simulated these proposals using our
estimated model for Army enlisted personnel to illustrate the implications of these alternatives for retention, cost, and other outcomes. We
expect the results for the other services to be similar. The DACMC
proposals that we considered and their key retirement plan features are
listed in Table 5.1. The first proposal, "DC + DB annuity," consists of
a DC plan, such as the Thrift Savings Plan for federal workers, plus a

DB plan. Under the DC plan, each member has an investment fund to which DoD contributes each year an amount equal to 5 percent of annual basic pay (ABP). Members may contribute to the funds, but their contributions are discretionary and are not included in our analysis; the analysis focuses on the DoD contributions. The DC plan vests at YOS 10 and begins payment at age 60. We assume that members convert the funds to an annuity and receive the annuity until age 100, and we discount the annuity back to age 60 using the personal discount rate and survival probabilities for each age. The DB plan has the same formula as the current high-three plan. The formula is 2.5 percent times a member's highest three years of basic pay, times YOS. The DB plan begins benefit payments at age 60, and in the first proposal, it vests at YOS 10.

The second proposal, "separation pay," includes the DC plan, with the DB plan vested at YOS 20, rather than YOS 10, plus separation pay. Separation pay is payable in the year the individual leaves service, assuming that the individual has completed 10 years of service. The payment is equal to a multiplier M times monthly basic pay (MBP) times years of service. Under the second proposal, M equals 0.50.

The third proposal, "gate pay," consists of the DC plan, with the DB plan vested at YOS 20, plus a series of payments called *gate pay* that are paid to those who complete certain milestones. Specifically, the service member receives a gate pay of 50 percent of ABP upon completion of YOS 10, 15, 25, and 30. The gate payment is not conditional on continuation, so a member who completes 10 years of service and separates receives the same payment as one who completes 10 years and stays. The fourth proposal, "gate pay + early vesting," is identical to the third proposal, except that the DB plan is vested at YOS 10 rather than YOS 20.

The fifth proposal, "hybrid," is identical to the second proposal but includes gate pay. That is, this proposal consists of the DC plan, the DB plan vested at YOS 20, separation pay with a multiplier of 0.75, and gate pay. The final proposal, "FERS," resembles the FERS plan. The plan includes the DC plan with a higher DoD annual contribution—10 percent of ABP rather than 5 percent. It also includes a DB plan for which

Table 5.1
DACMC Proposal Features

DACMC Option	DC Plan		Separation Pay		Immediate Annuity		DB Plan		Gate Pay
	Value	Vest (years)	Value	Vest (years)	Value	Vest (years)	Value	Vest (years)	
Current					2.5% × high-3 pay × YOS	20			
1. DC + DB plans	5% × ABP	10					2.5% × high-3 pay × YOS	10	
2. Separation pay	5% × ABP	10	0.5 × MBP × YOS	10			2.5% × high-3 pay × YOS	20	
3. Gate pay	5% × ABP	10					2.5% × high-3 pay × YOS	20	50% × ABP at YOS 10, 15, 20, 25, and 30
4. Gate pay + early vesting	5% × ABP	10					2.5% × high-3 pay × YOS	10	50% × ABP at YOS 10, 15, 20, 25, and 30
5. Hybrid	5% × ABP	10	0.75 × MBP × YOS	10			2.5% × high-3 pay × YOS	20	50% × ABP at YOS 10, 15, 20, 25, and 30
6. FERS	10% × ABP	10					1.5% × high-3 pay × YOS	10	75% × ABP at YOS 10, 15, 20, 25, and 30

the formula is 1.5 percent rather than 2.5 percent of ABP times YOS, with vesting at YOS 10. Finally, this proposal also includes gate pay.

Table 5.2 presents some summary statistics resulting from simulations of these proposals, as well as the current system (i.e., the base case for the Army). Figure 5.1 shows that, for the Army, the first proposal, an "old-age" plan that pays DC and DB benefits beginning at age 60, reduces midcareer retention. The percentage of individuals who stay until YOS 10 declines from 22.4 percent in the base case to 22.0, and the percentage staying until YOS 20 declines from 10.5 percent to 6.5 percent. Expected man-years per accession decline to 6.72 from 7.00, implying that accessions must increase to 61,771 to maintain Army strength at 415,000. Not surprisingly, active-duty cost per man-year falls as well, to $41,531. Reserve cost per man-year increases to $9,270.

Adding gate pay to the DC and DB plans, the third proposal, restores midcareer retention, as shown in Figure 5.2, and increases retention after YOS 20 relative to the base case. Under the proposal, 9.0 percent reach YOS 20, and expected man-years per accession increase

Table 5.2
Army Active-Duty Outcomes, Current Base-Case and DACMC Proposals

DACMC Option	Expected Active Years of Service per Accession	Average Active-Duty Cost per Man-Year ($)	Total Active-Duty Cost ($ millions)
Current	7.0	46,346	19,234
1. DC + DB plans	6.7	41,531	17,235
2. Separation pay	6.7	39,892	16,555
3. Gate pay	7.4	44,018	18,267
4. Gate pay + early vesting	7.4	44,777	18,582
5. Hybrid	7.7	44,880	18,625
6. FERS	7.8	44,738	18,566

NOTE: Simulations based on 5,000 cases. Assumed strength under all proposals is 415,0000.

Figure 5.1
**Simulated Active-Duty Members, by YOS, of Enlisted Army Personnel Who
Entered Active Duty: DACMC "DC + DB" Proposal**

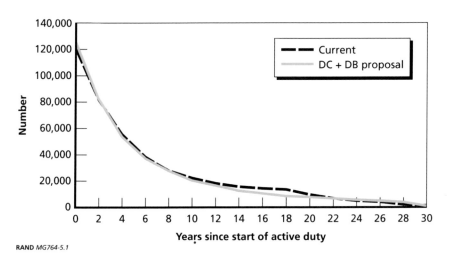

RAND MG764-5.1

Figure 5.2
**Simulated Active-Duty Members, by YOS, of Enlisted Army Personnel Who
Entered Active Duty: DACMC "DC + DB + Gate Pay" Proposal**

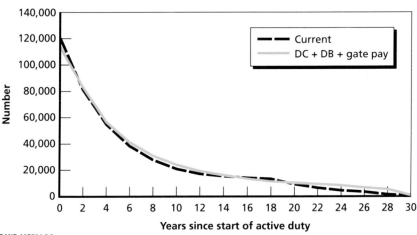

RAND MG764-5.2

to 7.4. The increase in active-duty man-years per accession and retention is accompanied by a decrease in active-duty cost per man-year, to
$44,018.

Earlier vesting of the annuity, the fourth proposal, has similar effects to those of third. Under this proposal, 10.3 percent reach
YOS 20, the same as under the current system, and expected man-
years per accession increase to 7.4 years (see Figure 5.3). Active-duty
cost per man-year, $44,777, is about the same as under the current
system. Thus, the third and fourth proposals are more cost-effective
than the current system, because they provide greater retention at the
same or lower cost.

The sixth proposal, "FERS," also improves midcareer retention
and results in longer careers after YOS 20, as shown in Figure 5.4. Furthermore, midcareer retention increases between YOS 4 and 12. Under
this proposal, 26.6 percent reach YOS 10, and 10.3 percent reach YOS
20. This proposal results in a more cost-effective system, because average man-years per accession increase to 7.8, while the cost per man-
year, $44,738, remains about the same as under the current system.

Figure 5.3
**Simulated Active-Duty Members, by YOS, of Enlisted Army Personnel Who
Entered Active Duty: DACMC "DC + DB + Gate Pay" Proposal and "DC + DB
+ Gate Pay (Early Vesting)" Proposal**

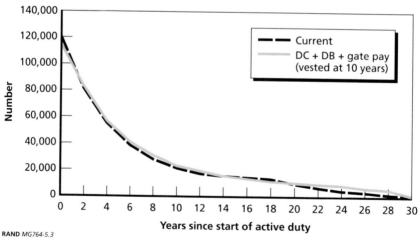

**Figure 5.4
Simulated Active-Duty Members, by YOS, of Enlisted Army Personnel Who
Entered Active Duty: DACMC "FERS" Proposal**

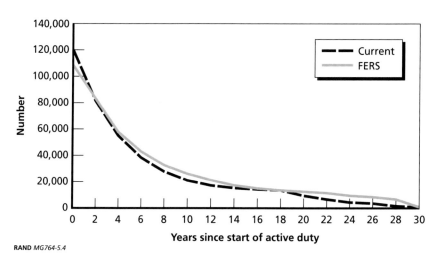

Finally, the addition of separation pay, vested at YOS 10, induces shorter careers. As shown in Figure 5.5 for the "separation pay" proposal, midcareer retention declines as members opt to take separation pay. Active man-years per accession decrease, as does active cost per man-year. However, adding gate pay (the "hybrid" proposal) restores midcareer retention and, indeed, increases retention among those with six to eight years of service. Under the "hybrid" proposal (see Figure 5.6), active man-years per accession increase relative to the current system.

As discussed by the DACMC, the various proposals address the key criticisms of the current retirement system. They are more cost-effective because, depending on the value of gate pay, they provide the same or more retention at lower cost. Furthermore, they are fairer to junior members, because vesting under the DC plan and, in some cases, under the DB plan, occurs before YOS 20, so more personnel become vested. By allowing the timing and amount of separation pay and gate pay to vary, the proposals also provide force managers with greater flexibility to tailor careers to meet changing requirements. For example, those in "youth-and-vigor" occupations could have shorter careers, as

Figure 5.5
**Simulated Active-Duty Members, by YOS, of Enlisted Army Personnel Who
Entered Active Duty: DACMC "Separation Pay" Proposal**

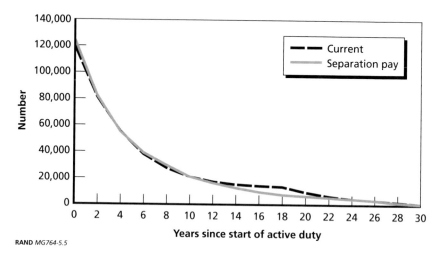

RAND *MG764-5.5*

Figure 5.6
**Simulated Active-Duty Members, by YOS, of Enlisted Army Personnel Who
Entered Active Duty: DACMC "Hybrid" Proposal**

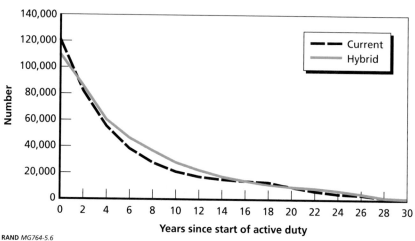

RAND *MG764-5.6*

illustrated for the "separation pay" proposal. Longer careers could be achieved as well, as illustrated for the "gate pay" proposal. While the DACMC did not propose a specific plan, its proposals provide a basis for the development of such a plan that addresses past criticisms of the current system.

QRMC Proposal

The QRMC developed a retirement system proposal, informed by the results of the DACMC options. The features of the QRMC proposal are presented in Table 5.3. The DC and DB plans are the same across the services, but the values of gate and separation pay may differ by service. The gate pay and separation pay values shown are for the Army, and the values for the other services are given later when we discuss their results. The specific gate pay and separation pay values were chosen to produce certain active-duty profiles, based on guidance from the QRMC director. These are described in more detail later in this section.

 Like the DACMC proposals, the QRMC proposal includes a DC plan, a DB plan, separation pay, and gate pay. The DC plan vests at YOS 10 and pays out at age 60. As shown in Table 5.3, the DoD contribution rate, as a percent of ABP, would vary by YOS, with a maximum rate of 5 percent for those with 10 or more years of service. The DB plan vests at YOS 10 and provides a benefit, payable beginning at age 60, equal to 2.5 percent times high-three ABP times YOS. For those with 20 or more years of service, the annuity would be payable at age 57, using the same formula. The QRMC proposal would offer members with 20 or more years of service an early-withdrawal option. A member with 20 or more years of service can take an immediate but reduced annuity. The annuity is reduced by 5 percentage points for each age at which the member is younger than 57. Thus, a 45-year-old member with 25 years of service could choose between waiting until age 57 and getting the full annuity of $2.5\% \times \text{high-3} \times \text{YOS}$ or beginning the payout of the annuity immediately upon separation from service at age 45, with the annuity reduced by 12×0.05, or

Table 5.3
QRMC Proposal Features for the Army

QRMC Option	DC Plan		Separation Pay		DB Plan		DB Plan Early-Withdrawal Option		Gate Pay
	Value	Vest	Value	Vest	Value	Vest	Value	Vest	
QRMC current	0% × ABP 0≤YOS≤1 2% × ABP YOS = 2 3% × ABP YOS = 3 4% × ABP YOS = 4 5% × ABP YOS 5+	10	1.0 × MBP × YOS	20	2.5% × high-3 ABP × YOS	10	Reduce annuity by 0.05 × (57–age)	20	15% × ABP at YOS 12 and 18
QRMC long	0% × ABP 0≤YOS≤1 2% × ABP YOS = 2 3% × ABP YOS = 3 4% × ABP YOS = 4 5% × ABP YOS 5+	10	1.0 × MBP × YOS	20	2.5% × high-3 ABP × YOS	10	Reduce annuity by 0.05 × (57–age)	20	25% × ABP at YOS 12 35% × ABP at YOS 18

60 percentage points. In other words, the member would get 40 percent of the $2.5\% \times \text{high-3} \times \text{YOS}$ annuity that he or she would have gotten at age 57.

The proposal also includes gate pay and separation pay. The value and timing of gate and separation pay will depend on the career length and retention profile that the service wishes to achieve. In the following analysis, we focus on two profiles, based on QRMC guidance. The first is the retention profile resulting from the current system, shown in Figure 4.1 in Chapter Four. To achieve this profile, gate pay is 15 percent of ABP and paid to those reaching YOS 12 and 18, while separation pay is payable to those with 20 or more years of service, and the multiplier is 1.0 (labeled "QRMC current" in Table 5.3). In the analysis, we find that providing separation pay to those with 20 to 24 years of service is sufficient to achieve the current Army profile, given the other components of compensation. The second is a retention profile that results in somewhat longer Army careers. This profile would increase retention beyond YOS 20, relative to the current profile, as well as increase midcareer retention slightly. Achieving this profile requires setting gate pay at YOS 12 and 18 equal to 15 percent and 35 percent, respectively, and the separation multiplier to 1.0 (labeled "QRMC long" in Table 5.3). Separation pay is provided to those with 20 to 30 years of service.

These two QRMC profiles show not only that the QRMC proposal can replicate the current force structure or produce a force structure with longer careers, but they also implicitly demonstrate that different retention profiles can be produced for different communities within a service by means of setting different gate pays and separation pays for those communities. Thus, the results support the proposition that the QRMC proposal offers flexibility in creating careers of varied lengths within a service and in varying the length of a career over time as the requirements vary for personnel in a community.

Figure 5.7 shows the simulated active-duty retention profile, given our estimates, under the current system and under the "QRMC Current" proposal. As is clear from the figure, the QRMC proposal is able to replicate the current force profile through the judicious setting of gate pay and separation pay.

Figure 5.7
Simulated Active-Duty Members, by YOS, of Enlisted Army Personnel Who Entered Active Duty: QRMC Current Proposal

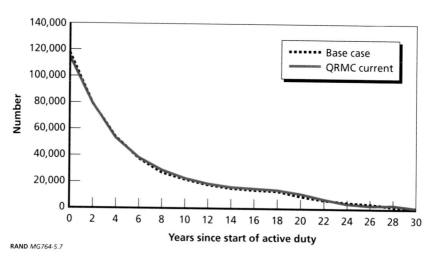

Table 5.4 shows active-duty cost per man-year, as well as other summary statistics of the QRMC options. The table shows that the "QRMC Current" proposal is less costly than the current system. Active-duty cost per man-year falls to $43,168 from $46,346, or 6.9 percent. Thus, this proposal achieves the current force profile more cost-effectively. Costs decline because this system is more front-loaded than the current system. That is, a higher percentage of lifetime wealth is incurred earlier in the career. Given that members have a higher personal discount rate than the rate at which the government discounts costs, pay that occurs earlier in the career is valued more than wealth that is paid later. Thus, by moving compensation forward, the proposal achieves the same retention and about the same number of active-duty accessions, but at lower` cost. The amount of the cost savings will depend on the personal discount rate. We use 15 percent because this rate provides a reasonable fit for the active-duty force profile, as demonstrated earlier. However, this rate is assumed for all members, and individual rates may differ from the one that fits the data. Rates that are lower and closer to the government discount rate will result in small cost

Table 5.4
Active-Duty Outcomes, Current Base-Case and QRMC Proposals

Proposal	% Reaching YOS 20	Expected Active Years of Service per Accession	Expected Active- Duty Accessions	Average Active- Duty Cost per Man- Year ($)	Assumed Strength	Total Active- Duty Cost ($ millions)
Army						
Base case	10.5	7.0	59,483	46,346	415,000	19,234
QRMC current	10.8	7.1	58,283	43,168	415,000	17,914
QRMC long	12.5	7.6	54,964	45,839	415,000	19,023
Air Force						
Base case	22.4	9.5	29,380	52,873	280,000	14,805
QRMC current	23.6	10.1	27,829	49,565	280,000	13,878
QRMC long	27.0	10.3	27,106	51,262	280,000	14,353
Navy						
Base case	14.8	7.7	29,118	49,194	225,000	11,069
QRMC current	12.6	7.7	29,360	44,503	225,000	10,013
QRMC long	15.9	8.4	26,906	47,925	225,000	10,783
Marine Corps						
Base case	10.0	6.7	26,813	46,780	180,000	8,420
QRMC current	10.1	7.0	25,798	43,920	180,000	7,906
QRMC long	12.7	7.7	23,246	46,149	180,000	8,307

NOTE: Simulations based on 5,000 cases.

savings or possibly no cost savings, while personal discount rates that are higher will increase the cost savings.

To produce somewhat longer careers with more midcareer retention and fewer accessions and to achieve the same strength, the separation pay multiplier is increased, as is the amount of gate pay. Figure 5.8 shows the simulation results for the "QRMC Long" proposal. More personnel stay beyond YOS 20, as well as through midcareer.

Figure 5.8
Simulated Active-Duty Members, by YOS, of Enlisted Army Personnel Who Entered Active Duty: QRMC Long Proposal

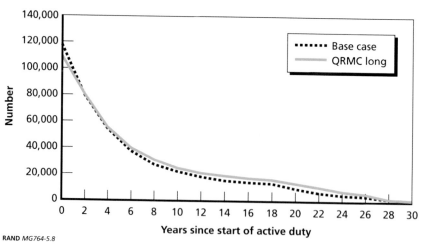

Man-years per accession increase to 7.6, or by 9 percent in the Army, and active-duty accessions fall from 59,483 to 54,964. Table 5.4 reports that the percentage of personnel who reach YOS 20 is slightly higher as well. However, active-duty cost per man-year is about the same or less. Thus, under the QRMC proposal, greater retention can be achieved at the same or lower cost, implying that this proposal is more cost-effective than the current system. Again, the reason is the same as before: A larger fraction of compensation is paid earlier in the career, in the form of gate pay and separation pay, and a smaller fraction is paid later, in the form of retirement pay. Given relatively high personal discount rates, this shift results in a more cost-effective system.

The differences between the current and the long-career versions of the QRMC proposal are not large and reflect interest by the QRMC in the possibility of achieving modest increases in retention prior to YOS 20 and after YOS 20. To illustrate the capability of the QRMC proposal to generate even more significant departures from the current profile through the appropriate use of gate pay and separation pay, Figure 5.9 shows, for the Army case, a version of the QRMC proposal that produces a shorter career and one that produces a longer

Figure 5.9
Simulated Active-Duty Members, by YOS, of Enlisted Army Personnel Who Entered Active Duty: QRMC "Shorter" and "Longer" Proposals

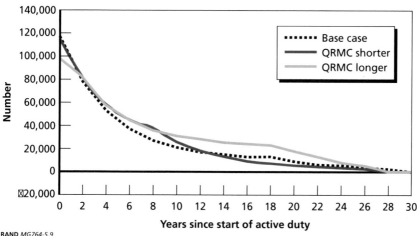

RAND *MG764-5.9*

career than the one shown in Figure 5.8. In the shorter-career case, gate pays are eliminated, and separation pay, equal to 1.75 times MBP times YOS, is paid to those who have at least 10 years of service. In the longer-career case, gate pays are set to 40 percent of ABP and paid to those reaching YOS 12, 14, 16, or 18. Separation pay is vested at YOS 20 and also equals 1.75 times MBP times YOS. As the figure shows, the shorter-career version induces more individuals to stay until YOS 10, when separation pay and the DB plan are vested, but fewer to stay until YOS 20. The longer-career version induces greater retention across years of service.

Force managers in the services might be interested in using the shorter-career version to target specific communities in which a short career is desired, such as combat arms, and in using the longer-career version to target other communities in which a longer career is desired. To consider the cost implications of such an approach, we arbitrarily assume that the Army wants one-third of its force to have a short career, one-third to have the current career, and one-third to have a longer career. Table 5.5 shows the cost per active man-year under the three versions and the weighted average (using 1/3, 1/3, and 1/3 as

Table 5.5
Active-Duty Outcomes, Army Current Base-Case and QRMC Proposals

Proposal	% Reaching YOS 20	Expected Active Years of Service per Accession	Expected Active-Duty Accessions	Average Active Duty Cost per Man-Year ($)	Total Active-Duty Cost ($ billions)
Base case	10.5	7.0	59,483	46,346	19.234
QRMC current	10.8	7.1	58,283	43,168	17.914
QRMC long	12.5	7.6	54,964	45,839	19.023
QRMC shorter	31.1	7.1	58,309	43,560	18.077
QRMC longer	29.8	8.5	48,572	50,388	20.911
Weighted average of base case, QRMC longer, and QRMC shorter	27.5	7.5	54,900	46,297	19.213

NOTE: Simulations based on 5,000 cases. Assumed strength for all proposals is 415,000.

weights). The table indicates that the shorter-career version costs less, and the longer-career version costs more, than the current system, but the weighted average costs about the same or slightly less than the current system. Specifically, the current system costs $46,346 per active man-year, and the weighted average is $46,297 per active man-year. These results suggest that the Army could achieve more variation in career lengths for about the same cost as the current system and while continuing to allow personnel to choose how long they prefer to stay.

Figures 5.10 and 5.11 show the simulation results of the QRMC Current and QRMC Long proposals for the Navy. Figures 5.12 through 5.15 show them for the Air Force and Marine Corps, respectively.[1]

---

[1]   For the Navy, the gate pay multipliers are 15 percent and 35 percent, respectively, at YOS 12 and 18, and the separation multiplier is 1.0 for QRMC Current; the gate pay multipliers are 25 percent and 35 percent, respectively, at YOS 12 and 18, and the separation multiplier is 1.25 for QRMC Long. For the Air Force, the gate pay multipliers are 15 percent and 15 percent, respectively, at YOS 12 and 18, and the separation multiplier is 1.25 for QRMC Current; the gate pay multipliers are 25 percent and 25 percent, respectively, at YOS 12 and 18, and the separation multiplier is 1.75 for QRMC Long. For the Marine Corps, the

Summary statistics are shown in Table 5.4, earlier in this chapter. As for the Army, the QRMC Current proposal achieves the current retention profile and similar accessions at lower cost for the other services. Specifically, active-duty cost per man-year drops by 6.3 percent, 9.5 percent, and 6.1 percent for the Air Force, Navy, and Marine Corps, respectively, relative to the base case, while active-duty man-years per accession are virtually unchanged, and accessions change little. Also, as for the Army, the QRMC Long proposal achieves longer careers in the other services, at about the same or lower cost. Active man-years per accession increase by 8.4 percent, 9.1 percent, and 15.3 percent, while active cost per man-year falls by 3.1 percent, 2.6 percent, and 1.3 percent in the Air Force, Navy, and Marine Corps, respectively. Active-duty accessions also decline, because, with higher retention, fewer accessions are needed to sustain the assumed strength levels.

---

gate pay multipliers are 10 percent and 10 percent, respectively, at YOS 12 and 18, and the separation multiplier is 0.75 for QRMC Current; the gate pay multipliers are 15 percent and 30 percent, respectively, at YOS 12 and 18, and the separation multiplier is 0.75 for QRMC Long. Under the QRMC Current proposal, separation pay is provided to those leaving between YOS 20 and 24. Under the QRMC Long proposal, separation is paid to those leaving between YOS 20 and 30.

Figure 5.10
Simulated Active-Duty Members, by YOS, of Enlisted Navy Personnel Who Entered Active Duty: QRMC Current Proposal

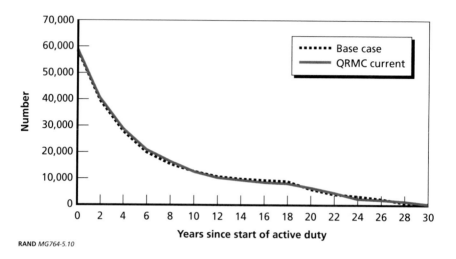

RAND MG764-5.10

Figure 5.11
Simulated Active-Duty Members, by YOS, of Enlisted Navy Personnel Who Entered Active Duty: QRMC Long Proposal

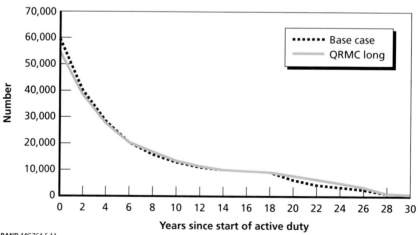

RAND MG764-5.11

Figure 5.12
Simulated Active-Duty Members, by YOS, of Enlisted Air Force Personnel
Who Entered Active Duty: QRMC Current Proposal

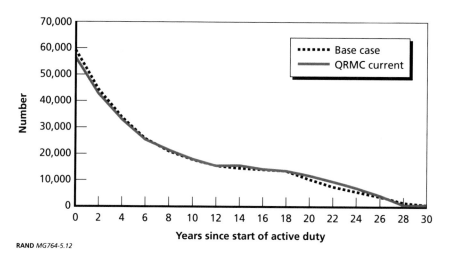

RAND MG764-5.12

Figure 5.13
Simulated Active-Duty Members, by YOS, of Enlisted Air Force Personnel
Who Entered Active Duty: QRMC Long Proposal

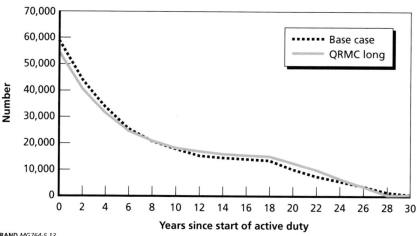

RAND MG764-5.13

Figure 5.14
Simulated Active-Duty Members, by YOS, of Enlisted Marine Corps Personnel Who Entered Active Duty: QRMC Current Proposal

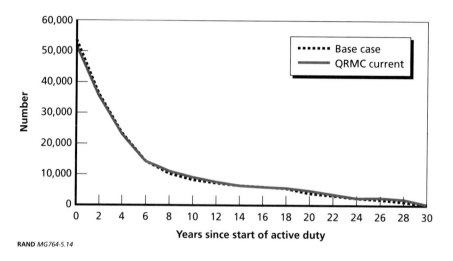

RAND MG764-5.14

Figure 5.15
Simulated Active-Duty Members, by YOS, of Enlisted Marine Corps Personnel Who Entered Active Duty: QRMC Long Proposal

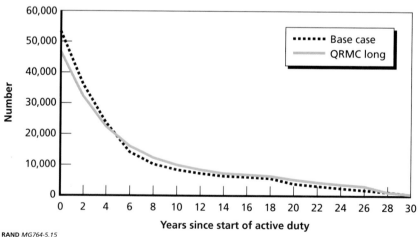

RAND MG764-5.15

CHAPTER SIX

Conclusions and Policy Implications

The QRMC proposal represents a major change in the structure of military compensation. This chapter summarizes our findings and discusses the proposal, drawing from the results presented in the previous chapter, in terms of how it addresses the criticisms of the current compensation system.

Summary of Findings from Simulations

The simulations of the policy alternatives under consideration by the QRMC led to the following findings:

- The policy alternatives could replicate each service's current active-duty retention profile. This required adapting the amount and timing of gate pay and separation pay for each service, but the DC and DB plans were the same across the services. Gate pay and separation pay are implemented at the discretion of each service, whereas it is expected that the DC and DB plans would be the same across the services. The latter is in keeping with the idea of fairness that is present in the current compensation structure, in which the tables for regular military compensation and retirement benefits are the same across the services but bonuses and special pays differ by service.
- The policy alternatives can shape the force, i.e., increase expected years of service, increase retention in the early career, and increase retention in the late career. Because shaping is accomplished

through gate and separation pays, a service could set these pays to achieve different retention profiles in different occupational areas and communities (e.g., special forces, logistics, linguistics). These profiles can also be changed over time in response to changes in manpower requirements by varying the level of the gate and separation pays in a community.

- The policy alternatives have the potential for cost savings. This depends on the exact terms of the compensation structure, and the alternatives examined have shown cost savings of 3–6 percent for the Army.

- The introduction of a DC plan as a major element of military compensation would change federal outlays (actual expenditures). Under the DC plan considered, a service member would vest after completing 10 years of service and would be entitled to the DC funds contributed by DoD on his or her behalf up to that point. That is, DoD, or perhaps the U.S. Treasury, would register an outlay equal to the amount in the service member's DC account at the time of vesting, plus an additional outlay for contributions in each subsequent YOS. This will increase outlays as a result of the DC plan, because, heretofore, the retirement-benefit system did not generate outlays for accruing liabilities but instead paid an accrual charge to the U.S. Treasury that was an intragovernmental transfer, not an outlay. On the other hand, government outlays in terms of the payment of annuities to retirees under the DB plan will decrease because the payment is deferred until age 57 or age 60, except for those who take a substantially reduced annuity immediately.

- Under the policy alternatives, a higher fraction of service members vest for retirement benefits than under the current system. The alternatives vest the DB plan at 10 years of service, not 20. The percentage vesting depends on the terms of the alternative, but in the simulations, the number vesting increased from 10.5 percent at YOS 20 to 23.7 percent at YOS 10—more than double.

- The alternatives succeed in creating an incentive to bring more service members to 10 years of service. This is expected to increase retention among junior enlisted and officer personnel, which can increase their expected years of service and, hence, their experience level and expected productivity.

- The alternatives could remove the strong incentive to leave active duty after completing 20 years of service. As pointed out by the DACMC, in certain occupations, such as "health professions, law, languages, cryptology, engineering, information technology, and other technical and scientific occupations" (DoD, 2006, p. xix), careers that extend beyond 20 years may be useful. More generally, in any field, there may be "master craftsmen," whose breadth and depth of knowledge and skills makes them highly attractive to retain for longer careers.

- The alternatives would affect the amount and timing of retirement-benefit payments available to vested service members. For those vesting in an active component, for example, the current system pays 50–75 percent of basic pay from the year of retirement from the military until the end of life. The alternatives offer higher current compensation in the form of gate and separation pays but would not begin benefit payments until age 60 for the DB plan (age 57 for those with 20 or more years of service) and age 60 for the DC plan. However, the member could opt to receive the DB annuity at a reduced rate immediately after leaving with 20 or more years of service. The option to take an immediate, reduced annuity, coupled with the DC plan, gate pay, and separation pay, can be attractive to service members. In particular, service members would have a stronger incentive to have a longer career—one extending beyond 20 years of active duty, as we illustrate next.

Compensation from the Perspective of the Member Leaving Active Duty

One way of relating the QRMC proposal to the individual service member is to compute the expected amount of money available to service members who leave at different years of service under the current system and under the alternatives suggested by the QRMC. This comparison of the actuarial value of military compensation for those who leave at different years of service provides information on whether members are made better off by the reform. We show these computations in Table 6.1. To make the computation, we assume that the basic-pay table, allowances, bonuses, special pays, and so forth, will be the same under the QRMC alternatives as under the current system. As a result, our computation focuses on a comparison of the present values at the time of leaving of (1) the defined benefits under the current system—that is, the present value of the retirement-benefit payments—and (2) the present value of the payments available under the QRMC alternatives. These include the present value of the DB retirement annuity, the present value of the DC plan—that is, the value of DC contributions made by DoD on behalf the member, plus interest—plus the present value of gate and separation pays. The current system offers a DB plan with vesting at 20 years, which is payable immediately upon separation from service. The QRMC alternatives offer a DB plan with vesting at 10 years, payable at age 60, or at age 57 for those with 20 or more years of service, at the same rate as under the current system, i.e., a rate computed by the same formula, or a DB payable immediately at a reduced rate to those with 20 or more years of active duty. The QRMC alternatives also include a DC plan, vesting at 10 years of service, and the amount contributed by DoD to the service member's account, plus interest, would be available to the departing service member who has vested. We assume that gate pay (plus interest, assuming that they were

Table 6.1
FY 2004 High-Three Monthly and Annual Base Pay for Selected Active-Duty Service Members

Base Pay ($)	E-5 at YOS 10	E-7 at YOS 20	E-8 at YOS 24	E-9 at YOS 30
Monthly	2,340	3,342	4,081	5,055
Annual	28,076	40,100	48,974	60,656
	O-3 at YOS 10	O-5 at YOS 20	O-6 at YOS 24	O-6 at YOS 30
Monthly	4,569	6,563	7,898	8,285
Annual	54,824	78,761	94,774	99,425

deposited) would also be available to the leaving service member, as would separation pay for those serving at least 20 years. In the plans we analyzed, the amount of separation pay equals MBP at time of departure times years of service, and this is payable to those leaving with between 20 and 26 years of service.

Our examples include the departures of an E-5 at YOS 10, E-7 at YOS 20, E-8 at YOS 24, E-9 at YOS 30, O-3 at YOS 10, O-5 at YOS 20, O-6 at YOS 24, and an O-6 at YOS 30. Table 6.1 provides context for the comparisons by showing the 2004 high-three base pay for these individuals, which we use in our calculations.[1] In particular, high-three base pay at time of leaving is related to the DB calculation under the current system and the QRMC alternatives. Base pay at time of leaving is also used in determining the amount of separation pay.

Table 6.2 consists of two panels. The upper panel refers to the current retirement-benefit system and the lower panel refers to the QRMC alternatives, with most attention on the immediate payment option. Under the current system, the vested member receives retirement-benefit payments immediately upon leaving an active component, and the benefits are paid for the remainder of the individual's life. We have divided the period of benefit payment into two phases: benefits paid up to age 57 and benefits paid from age 57 for the remainder of the actuarially expected lifetime. We use age 57 because it is the age at which benefits would begin under the QRMC alternative if the member did not opt for immediate payment and had at least 20 years of service. Age 57 is also the age at which FERS benefits begin, so it is a point of comparison with FERS, which serves federal civilian employees.

Importantly, because the comparison is made from the perspective of the individual, values are discounted with a personal discount rate. In estimating our model, we found some difference across services in the personal discount factor that produced the best fit. It was 0.87 for the Army, 0.91 for the Navy, 0.88 for the Air Force, and 0.85 for the Marine Corps. These correspond to personal discount rates of 14.9 percent, 9.9 percent, 13.6 percent, and 17.6 percent, respectively,

---

[1] Base pay in 2008, adjusted for inflation, is somewhat higher than base pay in 2004, but it is close enough for the results in Table 6.1 to be essentially the same.

Table 6.2
Present Value of Funds Available to Leaving Service Members, Current System, Option 1 Immediate, and Option 1 Deferred, Discount Rate = 15% (thousands of dollars)

	Option	E-5 at YOS 10	E-7 at YOS 20	E-8 at YOS 24	E-9 at YOS 30	O-3 at YOS 10	O-5 at YOS 20	O-6 at YOS 24	O-6 at YOS 30
Current system	Annuity payment	0	20	29	45	0	39	56	75
	PDV to 57	0	111	162	255	0	217	313	418
	PDV at 57	0	87	127	199	0	170	245	327
	PDV of PDV 57	0	8	21	75	0	16	40	123
	PDV of annuity	0	120	183	330	0	233	354	541
Option 1 Immediate	Annuity payment	0	3	10	30	0	6	20	48
	PDV to 57	0	21	65	142	0	40	126	232
	PDV at 57	0	21	71	207	0	41	137	340
	PDV of PDV 57	0	2	12	78	0	4	22	128
	PDV of annuity	0	23	78	227	0	45	150	372
	DC	11	40	57	93	22	76	108	169
	Separation pay	0	67	98	0	0	131	190	0
	Gate pay	0	9	10	12	0	20	23	28
	Total	11	138	243	333	22	271	470	569
	Total, if deferred	12	128	198	226	25	252	384	394

NOTE: Amounts have been rounded to the nearest $1,000.

for the Army, Navy, Air Force, and Marine Corps. We therefore conducted sensitivity analyses in our computation of the actuarial values and recomputed the figures in Table 6.2 using a discount factor of 0.90, implying a discount rate of about 10 percent, 0.889 (12.5 percent), and 0.95 (5 percent). The recomputed tables are shown in Appendix D.

Reading from the row of Table 6.2 labeled "PDV to 57," we find a PDV of $111,000 for retirement benefits paid to an E-7 retiring at YOS 20, which we assume to be age 40. The next row, "PDV at 57," shows that the PDV at age 57 of benefits paid from age 57 over the remainder of life equals $87,000. When this amount is discounted back to YOS 20, it equals $8,000. The major reason that it is not larger is expected mortality. Adding together the PDV at YOS 20 of benefits paid to age 57 and that of benefits paid thereafter, the total PDV of benefits is $120,000. Similar calculations show, for example, values of $183,000 for an E-8 leaving at YOS 24, $233,000 for an O-5 leaving at YOS 20, and $354,000 for an O-6 leaving at YOS 24.

Turning to the bottom portion of the table, we assume that the service member has chosen to take an immediate, reduced benefit. If he or she had waited until age 57 to begin receiving the benefit, the benefit amount would have been equal to the benefit under the current system. But opting to take the benefit immediately reduces it by a factor of 0.05 times (57 minus age at leaving). An E-7 leaving at YOS 20 and taking a benefit immediately receives a benefit that is $1 - 0.05 \times (57 - 40) = 0.15$ as large as the benefit under the current system. As seen in the row labeled "Annuity payment," the benefit is $3,000. This is 0.15 of the $20,000 under the current system. (Numbers in the table have been rounded to the nearest $1,000.) Using this benefit amount and doing PDV calculations, we find the PDV of this annuity to be $23,000. In addition to the immediate annuity, the E-7 has funds in his or her DC account worth $40,000. Gate pays received in earlier years (in our simulation, received at YOS 12 and 18) are worth $9,000 (payment plus interest, assuming that the gate pays were invested at 4 percent). Finally, the separation pay is $67,000. The grand total of benefits for the E-7 is $138,000, which is $18,000 more than under the current system, given the assumption of a 15-percent personal discount rate. In fact, for each of the example service members in Table 6.2, the QRMC immediate-payment option provides the leaving service member with more money than under the current system.

In real life, the personal discount rate may vary from person to person and might be higher for younger personnel and perhaps lower for personnel with more education. A higher personal discount rate would tend to make the QRMC proposal more attractive to individual members than shown in Table 6.2, because its compensation is less back-loaded than that in the current system, while a lower discount rate would tend to make the QRMC proposal less attractive. Examples are shown in Appendix D, Tables D.1 through D.3, in which we consider personal discount rates of 5 percent, 10 percent, and 12.5 percent. When the discount rate is 5 or 10 percent, we generally find the opposite of the results shown in Table 6.2. That is, the QRMC proposal is less attractive than the current system in terms of the wealth it provides to individuals for most of the cases that we considered for enlisted and officer personnel. When the discount rate is 12.5 percent, the QRMC proposal is more attractive to individual members than the current system, similar to the results in Table 6.2, except for those at YOS 30. Using data from the defense drawdown in the 1990s and members' choice between an annuity and lump-sum buyout to infer personal discount rates, Warner and Pleeter (2001) found that the mean nominal personal discount rate is 10.4 to 18.7 percent for officers (depending on the model estimated) and 35.4 percent to 53.6 percent for enlisted personnel. Personal discount rates varied with demographic characteristics, such as number of dependents, and job characteristics, such as YOS. For example, for officers, the mean personal discount rate was between 20.5 percent and 29.1 percent at YOS 7 but between 0 and 9.9 percent at YOS 15. In Warner and Pleeter's analysis, about half of the difference between officer and enlisted personal discount rates was due to differences in observable demographic characteristics and the value of the annuity versus lump-sum choice. Holding these factors constant (i.e., giving enlisted personnel the same demographic characteristics as officers and the same buy-out choice) the mean personal discount rate for enlisted is 17.3 percent, versus 10.4 percent for officers.

The implications of differences in personal discount rates and the results in Tables 6.2 and D.1, D.2, and D.3 is that, for many (and perhaps most) enlisted personnel and junior officers, among those for whom the personal discount rate exceeds 12.5 percent, the QRMC proposal is more attractive in terms of member wealth, except for those at YOS 30. But for personnel with discount rates below 12.5 percent, including midcareer and more senior officers, those with more educa- tion, and those with fewer dependents, the QRMC proposal may not represent an improvement in wealth.

Table 6.2 shows that the gain of the QRMC system over the cur- rent system is greatest for an E-8 at YOS 24 and an O-6 at YOS 24, assuming the 15-percent personal discount rate. This implies a stronger incentive to serve for more than 20 years, and we see this effect at work in the estimates of average years of service in Chapter Five, which show higher average years under the QRMC alternative than under the cur- rent system. The size of the incentive is at the discretion of the service. The calculations have assumed that separation pay would be provided to those leaving with 20 to 24 years of service, with a multiplier of 1, but the services could extend the range to YOS 30 and use a higher multiplier if they chose to do so. This could make the relative gain highest from staying at a later YOS, e.g., YOS 30.

Table 6.2 also shows a positive present value for service members leaving at YOS 10 under the QRMC alternative, whereas they receive nothing under the current system. Service members with fewer than 20 years of service cannot opt for the immediate, reduced annuity and have not received gate pay or separation pay. (These service members will receive an annuity beginning at age 57, as discussed later in this chapter.) Therefore, the total of the present value of $11,000 comes from the DC.

We have also done computations for the QRMC alternative with the DB beginning at age 57. As mentioned, the amount of the DB equals that under the current system. But it does not begin until age 57, and this reduces its present value from the perspective of the member leaving at, say, YOS 20, 26, or 30. In a number of cases, we find this QRMC DB option to be of lower value than the QRMC immediate-benefit option, even though the immediate benefit is reduced, as described earlier. However, our calculations for the QRMC deferred annuity should be regarded as tentative because they assume the same separation pay and gate pay as for the QRMC immediate annuity. It is possible that separation and gate pay would have to be increased to offset the fact that the PDV of the deferred benefit was lower than that of the immediate benefit, and, if so, the deferred-annuity alternative would be more expensive to DoD than the immediate-annuity alternative.

For service members who leave at YOS 10, the present value is higher under the QRMC deferred-annuity alternative than under the current system (see Table 6.2). Service members with 10 years of service vest for a defined benefit, receivable at age 60, and the present value of this benefit at YOS 10 is $1,000, as of age 57. Added to the $11,000 for the DC, the combined total is $12,000 under the deferred-annuity option (last row of Table 6.2).

Discussion

The QRMC proposal would significantly alter the structure of the military retirement system. A key concern regarding such a major shift is whether an alternative retirement system could produce the same experience mix of personnel as the current system. The results in the previous chapter show that the QRMC proposal would allow the services to maintain the current mix—or change it—and do so at the same or lower cost. As discussed in Chapter Two, cost is only one criticism of the current system, but it is a primary reason for reform. Other reasons include concerns about the equity of the current system and its stifling effect on force-management flexibility.

Equity

The issue of equity has several dimensions having to do with the vesting of benefits, as well as age of eligibility for receipt of benefits. The first is equity with respect to the civilian sector, in which vesting of retirement benefits is required by ERISA to occur no later than after five years of service, in contrast to the military's 20-year vesting feature. The second is equity between junior military members who are not eligible for retirement benefits if they separate and senior members with 20 or more years of service who are eligible. The third is equity between active-duty service members and those in the reserve components. Active and reserve members are covered by different military retirement systems. These systems differ in several respects, but perhaps the most contentious is the ability of active-duty members with 20 or more years of service to receive immediate benefits upon separation, whereas reservists with 20 or more years of creditable service must wait until age 60 to receive retirement benefits. An active-duty service member entering at age 20 would be eligible to receive benefits at age 40, 20 years earlier than a reservist who entered at age 20 and served for 20 years.

To address inequity with respect to civilian retirement plans and inequity between junior and senior military members, past studies and commissions have recommended earlier vesting of the military retirement benefit. However, as shown in the case of the DACMC "DC + DB" plan (in which both the DC and DB plans vest at YOS 10), earlier vesting alone does not necessarily have desirable force-management features or improve benefits for members. In the case of the DACMC plan, earlier vesting in a benefit paid at age 60 yields less generous retirement benefits than the current system and reduces retention. On the other hand, the QRMC proposal, a plan that couples earlier vesting with higher current pay in the form of gate and separation pays, can improve the wealth of members—or decrease it, depending on the gate-pay amounts and the member's personal discount rate—and can facilitate a variety of retention profiles, including the one resulting from the current system. It is more comparable with civilian plans because it vests earlier, at YOS 10 rather than at YOS 20, and because it includes a DC plan, a feature of civilian plans that has become quite common.

The QRMC proposal also does not distinguish between reserve and active status; the structure of the retirement system is the same for both active and reserve members, unlike the current system. That is, active and reserve members have the same vesting requirement and the same age of payout of benefits, age 60 for the DC plan and age 57 (for those with 20 or more years of service) or age 60 (for those with 10 years) for the DB plan. Thus, the proposal addresses this criticism of the current system and is consistent with the recommendation of the recent Commission on the National Guard and Reserves. However, because reservists serve less intensively, the value of their retirement benefit will be less than the value of the active benefit for the same number of years of creditable service.

Force Management

Concerns about the effects of the current system on force management stem from an understanding that the compensation system affects the ability of managers to influence service members' career outcomes. As discussed in Chapter Two, the current system produces similar retention patterns across occupations. Members at the midpoint in their careers, between YOS 10 and 20, are reluctant to leave and forgo the retirement benefit that they would receive if they stayed until YOS 20, and, understanding this behavior, the services are reluctant to separate any but the most marginal performers who have between 10 and 20 years of service, for fear of breaking the "implicit contract" between the service and midcareer members. Thus, the retirement system creates a type of "golden handcuff" that induces a similar profile of retention regardless of occupation.

Aside from inducing common career lengths, the retirement system hampers the flexibility of force management in other ways. Assignment lengths can become compressed, especially for officers and noncommissioned officers, as more assignments are packed into a 20-year career to ensure that members have the right number and mix of assignments to be effective as senior or noncommissioned officers. The frequent rotation of personnel means that individuals may have insufficient time to master all of the tasks in a given assignment and may take actions that focus on short-term over long-term results or that ignore the long-term consequences of current actions. The opportunity for a longer career, especially for personnel whom the military wants to keep, would enable longer assignments and less frequent rotation.

The QRMC proposal would allow for more variability in career length and, where longer career paths were chosen, enable less frequent rotation of personnel. Gate pay increases retention among junior personnel because there is increased incentive to stay and achieve the career milestone. Separation pay also increases retention among junior personnel, because the incentive to stay until the vesting point for separation pay increases. On the other hand, separation pay reduces retention after the vesting point (i.e., once members are eligible to receive separation pay), because it induces members to leave. Gate and separation pay could be added to the current compensation system and, presumably, would enable the services to obtain more varied career lengths than at present. But adding gate and separation pay to the current system would likely increase costs, whereas the QRMC proposal would decrease costs.

Cost-Effectiveness

The cost-effectiveness of the compensation system is improved if the same retention is achieved for less cost or if more retention is achieved at the same cost as the current system. As seen in Table 5.4 and Figures 5.7 through 5.15 in Chapter Five, the QRMC proposal improves cost-effectiveness. It can achieve the same active-duty retention profile as the current system for about 7-percent lower cost per man-year. It can also achieve long careers at lower cost or at about the same cost.

The sources of the cost savings are in shifting compensation toward higher current pay and less pay in the form of retirement pay. Retirement eligibility for a full annuity is age 60 (age 57 for those with 20 or more years of service under the DB plan). Although members have the choice of receiving an early annuity, the 5-percentage-point reduction factor for each age below 57 reduces the cost of the annuity for those who claim an immediate annuity, relative to the current system. Shifting compensation toward more current pay, in the form of gate pay and separation pay, reduces cost, because the personal discount rate is higher than the rate at which the government discounts future costs. Members do not value retirement pay as much as it costs to provide it. A dollar of pay that occurs earlier in the career in the form of RMC, gate pay, or separation pay is discounted for fewer years and is worth more to a junior member than a dollar of retirement pay that is paid at the end of the career.

Conclusion

Most importantly, the analysis presented in this monograph demonstrates the viability of the 10th QRMC compensation-reform proposal. The analysis finds that the proposed reform has the potential to reproduce the current personnel force structure at lower cost and add flexibility by enabling the services to lengthen or shorten careers at their discretion, while keeping with their assessment of appropriate experience and grade mixes by occupational area or community. Changing to a new compensation system is a challenging endeavor, and many questions remain about the advisability, benefits, costs, and timing of such a change. We hope that the analysis and findings of this study are a contribution to the ongoing policy debate and additional study.

The analysis also serves to illustrate the power of the dynamic retention model in providing empirical estimates of the effects on retention, retirement, and other key outcomes of major changes in the military compensation system. The focus of the analysis has been on assessing proposals offered by the DACMC and the 10th QRMC, but it is clear that the approach can be used to assess a range of compensation reforms and extended to other applications.

Description of Current Retirement Systems

There are currently three different systems under which active-duty retirement pay can be calculated. For members entering military service prior to September 8, 1980, active-duty retirement pay is computed using the formula in Equation A.1, and BP is simply basic pay on the date of separation. For members entering military service between September 8, 1980, and July 31, 1986, BP is calculated as the average of the highest 36 months (three years) of basic pay (high-three method). Under both of these systems, annual retirement pay is adjusted according to changes in the Consumer Price Index (CPI) urban wage-earners series.

$$Y = \left[0.50 + 0.025(YCS - 20)\right] \times BP, \qquad (A.1)$$

where YCS is years of creditable service.

Active-duty members who enter military service after July 31, 1986, must choose between two retirement systems in their 15th year of service. The first system is the high-three averaging system. The second system is known as *REDUX*. Under REDUX, active-duty members receive a $30,000 career retention bonus at 15 years of service. Their initial retirement pay is then calculated according to the following formula:

$$Y = \left[0.40 + 0.035(YCS - 20)\right] \times BP, \qquad (A.2)$$

where *BP* is calculated under high-three averaging. Between the year of retirement and age 62, retirement pay under REDUX is adjusted according to the CPI minus 1 percent. At age 62, REDUX makes two adjustments to retirement pay. The first is to adjust the multiplier to what it would have been had the member retired under the high-three averaging system. For example, a member retiring under REDUX with 20 years of service would receive 40 percent of *BP* between retirement and age 62 and 50 percent of *BP* thereafter. The second adjustment is to restore retirement pay to what it would have been had retirement pay been fully indexed to the CPI. Thus, at age 62, retirement pay is identical under REDUX and under the high-three averaging system. After age 62, however, retirement pay under REDUX is once again adjusted according to the CPI minus 1 percent.

Members of the reserve components who accumulate 20 qualifying years of service with the last eight years of qualifying service in the Ready Reserve are entitled to receive retirement pay beginning at age 60. The Ready Reserve encompasses the Selected Reserve, the Individual Ready Reserve, and Inactive National Guard. It excludes the Retired Reserve. A qualifying (or "creditable") year is a year of service in an active component or a year in a reserve component in which the individual accumulated at least 50 points (discussed next). Between October 1994 and September 2001, the number of qualifying years in the last years of reserve service was reduced from eight to six. No retirement pay is provided to members separating from the reserves with fewer than 20 qualifying years of service. Retired pay at age 60 is calculated based on years of creditable service when transferred from the Ready Reserve, and basic pay is calculated under one of several methods (discussed later in this appendix):

$$Y = YCS \times 0.025 \times BP, \qquad (A.3)$$

where *Y* is monthly retired pay, *YCS* is years of creditable service, and *BP* is MBP. Roughly speaking, years of creditable service are a prorated number of qualifying years of service. Specifically, years of creditable service are calculated by dividing a reservist's accumulated retirement points by 360. Retirement points are computed as follows:

- one point for each day of active-duty service
- one point for each period of inactive-duty training
- one point for each day in funeral honors duty status
- one point for each accredited three-credit hour correspondence course satisfactorily completed
- 15 points for each year of active-status membership in a reserve component.

Under current law, reservists may accumulate no more than 90 inactive-duty points (annual membership, inactive-duty training, and course-credit points) and a combined total of 365 active- and inactive-duty points in a single year. The restriction on inactive-duty points has been relaxed significantly in recent years. For retirement years prior to September 23, 1996, annual inactive-duty points were capped at 60. This limit increased to 75 points for retirement years between September 23, 1996, and October 30, 2000, and stands at 90 points after October 30, 2000. There is also a career limit on retirement points, 10,950, or 30 years of creditable service. A minimum of 50 points must be earned in a year for that year to count toward meeting the minimum of 20 calendar years of service for vesting in retirement pay. The average enlisted reservist separating from the Ready Reserve in FY 2000 had accumulated 2,984 retirement points over 25 calendar years of active-duty and reserve service. The average reserve officer retiring in FY 2000 had accumulated 3,585 retirement points over 27 calendar years of service. Median retirement-point accumulation among all reservists totaled 77 for enlisted members and 79 for officers in FY 2000.

The computation of *BP* depends on when the reserve member first entered military service and whether he or she transferred to the Retired Reserve upon separating from the Ready Reserve. For members entering prior to September 8, 1980, *BP* is the basic pay in effect for a given rank and calendar years of service when the member first begins to receive retirement pay. Importantly, a member can continue to accumulate calendar years of service (i.e., longevity) for the purposes of computing *BP* if he or she transfers to the Retired Reserve after separating from the Ready Reserve. Consequently, individuals who separate from the Ready Reserve prior to reaching the highest level of basic pay for a given rank can increase *BP* by remaining in the Retired Reserve. Members of the Retired Reserve are not required to participate in drilling or training but can be called to active duty without consent in the interest of national defense. They receive no compensation and do not accumulate retirement points.

For members who entered on or after September 8, 1980, *BP* is computed as the average of the highest 36 months of basic pay (high-three averaging). This system is known as the *high-3 system*. For reservists who transfer to the Retired Reserve, high-three averaging takes place over basic pay in their last three years of service in the Retired Reserve (typically, ages 57–59). For reservists who end their affiliation with the reserves upon separation from the Ready Reserve, *BP* is calculated over their last three years of service in the Ready Reserve. This distinction creates incentives for reservists to remain in the Retired Reserve until age 60, so that *BP* at age 60 reflects real wage growth subsequent to separation from the Ready Reserve, as well as any increases in pay due to changes in longevity. There is no incentive to delay retirement beyond age 60. All members below major general must separate by age 60, and limits on calendar years of service may force some reservists to separate before age 60. Retirement pay beginning at age 60 for all members is adjusted for inflation according to changes in the CPI for urban wage-earners.

There are a number of differences between the active and reserve retirement systems. The most significant difference is that active-duty members with 20 or more calendar years of service begin receiving retirement pay immediately upon separating from the active-duty force instead of at age 60, as under the reserve retirement system. Also, reserve retirement benefits depend on years of creditable service, which, in effect, are based on days of reserve service per year, whereas a year in an active component counts as a full year.

Estimation Method and Data Sources

Estimation Methods

The model has seven parameters: the mean active preference, mean reserve preference, variance of active preference, variance of reserve preference, covariance (or correlation) of active and reserve preferences, and the scale parameters τ and λ. We assume that the preferences have a normal distribution and that the random shocks have extreme-value distributions.

We could compute the value of the value function for every state in every period if we knew the values of an individual's active and reserve preferences, the random shocks that the individual drew, and the scale parameters. This would also require knowledge of the military pay scale, military retirement-benefit formula, and civilian wage curve; these are known and are inputs to the model. We also input the discount factor. (We discuss the discount factor later in this appendix, but for now, we mention that, when estimating the model, we try different values for the discount rate and use the one that results in the best fit.) Assuming that we had all this information, the computation would start from the end state and proceed recursively: at the end of period T, the last period of work life, the individual receives the present value of any military retirement benefits he or she may be owed. The value of these benefits depends on the individual's state in the last period, and the individual may be in any number of states, depending on active years, reserve years, and final pay grade. All possible states must be considered, because, from the viewpoint of the individual earlier in his or her career, he or she does not know what this final state will be. Given the values for the end states, the values for $T–1$ can be computed. For each possible state in $T–1$, the value of the value function for a given alternative in T depends on the pay in T for that alternative, the preference for it, and the expected value of the maximum in the next period, T. Using the values computed for $T–1$, we could apply the probability formulas derived here to obtain a quantitative value for the probability of each possible transition from each possible state. Further, using the values computed in T and $T–1$, we could compute the values for $T–2$, and so forth, back to the first period. Using these and the probability formulas, we could compute the entire set of transition probabilities for all periods. Finally, if we knew the career path that the individual actually traveled, we could string together the transition probabilities for each choice in that path and determine the likelihood of the path. Doing this for everyone in the sample would give us the sample likelihood.

Let P_i represent the probability of the individual's career, i.e., periods of active-duty service followed by the exact sequence of being a reservist and/or a civilian over a work life. In general, P_i depends on the individual's tastes for active and reserve service, and these may be expressed in terms of the parameters of the taste distribution. It also depends on the variance of the transitory shock. Hence,

$$P_i = P_i\left(\gamma_{ia}, \gamma_{ir}\right) = P_i\left(\gamma_{ia}, \gamma_{ir}; \mu_a, \mu_r, \sigma_a^2, \sigma_r^2, \rho, \tau, \lambda\right).$$

But the individual's preferences and random shocks are not known, so the above computation cannot be done on the basis of that knowledge. But given that our goal is to estimate the parameters of the model, which characterize the preference and shock distributions, we can make use of the assumptions about how the preferences and shocks are distributed. Although we (as analysts) do not know an individual's preferences, we have assumed that preferences are normally distributed. Therefore, whatever an individual's preferences, they are drawn from a normal distribution. Using this assumption, we can compute the expected value of the value function, where the expected value is taken over the taste parameters (i.e., the taste parameters are integrated out).

The expected value of the value function depends on the parameters of the preference distribution but not on the individual's own preferences. Similarly, we can again compute the expected transition probabilities and the likelihood of the career path—also no longer conditional on the specific values of the preferences:

$$\int\limits_{-\infty}^{\infty} \int\limits_{-\infty}^{\infty} P_i g\left(\gamma_a, \gamma_r\right) d\gamma_a d\gamma_r.$$

We do this for every individual, then multiply these together to create the likelihood function for the career paths of the individuals in the sample. We would like to maximize the likelihood function with respect to the parameters of the taste distribution and the scale parameters τ and λ. If the maximization can be done, even though the individual's tastes and shocks are not observed, the parameters of the taste distribution and shock distributions are estimable.

A difficulty with this approach is integrating out the preferences, because the integral for the normal does not have a closed form. To overcome this difficulty, we approximate the integral by evaluating it at 12 randomly drawn values for the active and reserve preferences, each, then take a simple average of the evaluations.

$$\int\limits_{-\infty}^{\infty} \int\limits_{-\infty}^{\infty} P_i g\left(\gamma_a, \gamma_r\right) d\gamma_a d\gamma_r \simeq \frac{1}{J} \sum_{j=1}^{J} P_{ij}\left(\gamma_{ija}, \gamma_{ijr}; \mu_a, \mu_r, \sigma_a^2, \sigma_r^2, \rho, \tau, \lambda\right).$$

Therefore, the simulated likelihood function is

$$SLL = \sum_{i=1}^{n} \log\left(\frac{1}{J} \sum_{j=1}^{J} P_{ij}\left(\gamma_{ija}, \gamma_{ijr}; \mu_a, \mu_r, \sigma_a^2, \sigma_r^2, \rho, \tau, \lambda\right)\right).$$

Evaluating the sum also creates a challenge: Parameter values must be specified for the normal distribution before the draws can be made, but these are the same parameter values that we want to estimate. The following approach provides a way out of this circle.

Following Train (2003), we take two independent draws from a standard normal distribution, $N(0, 1)$, and use a Cholesky decomposition to transform them into random variables that are jointly normally distributed with means $\mu = (\mu_1, \mu_2)$ and covariance matrix

$$\Omega = \begin{pmatrix} \sigma_{11}^2 & \sigma_{12}^2 \\ \sigma_{21}^2 & \sigma_{22}^2 \end{pmatrix}.$$

The covariance matrix is positive, definite, and symmetric, $(\sigma_{12}^2 = \sigma_{21}^2)$, so we can define a Cholesky matrix as follows:

$$L = \begin{pmatrix} \alpha_{11} & 0 \\ \alpha_{21} & \alpha_{22} \end{pmatrix},$$

such that $LL' = \Omega$.

In our case, let (η_{i1}, η_{i2}) be two draws from a standard normal for individual i and calculate $\gamma_i = \mu + L\eta_i$. The values $\gamma_i = (\gamma_{ia}, \gamma_{ir})$ follow a normal distribution because they are the sum of normals, which is normally distributed. The mean and covariance matrix of γ_i are

$$E(\gamma_i) = \mu + L E(\eta_i) = \mu$$
$$Var(\gamma_i) = E L\eta_i (\eta_i L)' = L E\eta_i \eta_i' L' = L Var(\eta_i) L' = LIL' = LL' = \Omega.$$

In our application,

$$\begin{pmatrix} \gamma_{ia} \\ \gamma_{ir} \end{pmatrix} = \begin{pmatrix} \mu_a \\ \mu_r \end{pmatrix} + \begin{pmatrix} \alpha_{11} & 0 \\ \alpha_{21} & \alpha_{22} \end{pmatrix} \begin{pmatrix} \eta_{i1} \\ \eta_{i2} \end{pmatrix} \text{ or } \begin{array}{l} \gamma_{ia} = \mu_a + \alpha_{11}\eta_{i1} \\ \gamma_{ir} = \mu_r + \alpha_{21}\eta_{i1} + \alpha_{22}\eta_{i2}. \end{array}$$

So, $Var(\gamma_{ia}) = \alpha_{11}^2$, $Var(\gamma_{ir}) = \alpha_{21}^2 + \alpha_{22}^2$, and $Cov(\gamma_{ia}, \gamma_{ir}) = \alpha_{11}\alpha_{21}$, which are the elements of the covariance matrix Ω.

As the estimation procedure iterates, the parameters $\mu_1, \mu_2, \alpha_{11}$, α_{21}, and α_{22} are updated, and these update the preferences γ_i via the

Cholesky equations. (The seed values, $\left[\eta_{i1}, \eta_{i2}\right]$, are not redrawn.) The shape parameters, τ and λ, also change. For a new set of estimates of preference distribution and shape parameters, the dynamic program must be resolved for each individual. The individual's new values are used to update the transition probabilities and the likelihood of the career path, and the sample likelihood is updated and remaximized with respect to the parameters. This continues until the parameter estimates converge on their final values; in other words, they change little from one iteration to the next.

The solution of the dynamic program is computer-intensive. We use Keane and Wolpin's (1994) interpolation method to shorten the computation time. Instead of recomputing the values of the value function, we substitute in interpolated values, resulting in much faster computation. Interpolation begins by computing the value of the value function for a grid of points in the state space consisting of several hundred thousand different combinations of active years, reserve years, pay grades, and years in the workforce. At each grid point, we compute the exact values of the value functions over a range of the parameter values that we prespecify. Using the values computed for each grid point, we estimate regressions for the value as a quadratic function of the means and the scale parameters, as suggested by Keane and Wolpin (1994). Then, in the estimation, we use the fitted regressions to predict the value as a function of the current estimates of the parameters. The interpolation regressions are

$$IV_t^a = b_0^a + b_1^a \mu_a + b_2^a \mu_r + b_3^a \tau + b_4^a \lambda + b_5^a \left(\mu_a\right)^2 + b_6^a \left(\mu_r\right)^2 + b_7^a \tau^2 + b_8^a \lambda^2 + \eta^a$$

$$IV_t^r = b_0^r + b_1^r \mu_a + b_2^r \mu_r + b_3^r \tau + b_4^r \lambda + b_5^r \left(\mu_a\right)^2 + b_6^r \left(\mu_r\right)^2 + b_7^r \tau^2 + b_8^r \lambda^2 + \eta^r$$

$$IV_t^c = b_0^c + b_1^c \mu_a + b_2^c \mu_r + b_3^c \tau + b_4^c \lambda + b_5^c \left(\mu_a\right)^2 + b_6^c \left(\mu_r\right)^2 + b_7^c \tau^2 + b_8^c \lambda^2 + \eta^c.$$

The error terms on the right side are assumed to be independent and identically distributed. The regressions are estimated by ordinary least squares.

Data Sources

The parameters of the model are estimated with WEX data from the DMDC, augmented with data from other sources. The WEX contains person-specific longitudinal records of active and reserve service. We use WEX data to track individual service or component and pay grade over time. Other data sources provide information on military pay, civilian wage, military promotion, and high-year-of-tenure rules.

Work Experience File

DMDC creates WEX data from the active-duty master file and the reserve-component common personnel data system file. DMDC uses these files to build a snapshot of all personnel for each reporting period. The WEX information that we used includes service and component, reserve category code (indicating whether the individual serves in the selected reserve), pay grade, and years of service computed from pay-entry base date. Our WEX analysis file includes longitudinal data for individuals who entered active duty between September 30, 1990, and December 31, 2007.

The time span of our data on active-duty entrants is 18 years. The file contains all entrants, so sample size is not an issue, but the window of observation is not long enough to include active-duty vesting in the current retirement system, which occurs when 20 years of active duty have been completed. This means that the model will be fit on active-duty retention decisions prior to vesting and active-duty retirement. However, the model can predict retirement behavior over the entire military career. The predictions rely on knowing the model parameters, which are estimated, and the civilian pay profile, military pay, and retirement-benefit schedules, as well as on the decisionmaking structure in the theoretical model. Although the absence of data on retention among those with 20 or more years of service does not prevent the estimation of the model, it would clearly be preferable to have actual data on retention at YOS 20 and later. In a few more years, the WEX data will extend to the point at which personnel reaching YOS 20 are included. The policy simulations presented in Chapter Four suggest that our model predicts retention reasonably well at YOS 20 and later. However, we should add that we tried to augment our data with data on service members who, at the start of our data window, were already in service. For instance, personnel with, say, 15 years of service in 1990 might be followed for 10 or more years. This would provide observations on retention at YOS 20 and several years beyond. To accommodate such data, we recognized that all individuals in an active component in 1990 are a selected sample conditional on "surviving" on active duty from their entry year to 1990. This means that their preference distribution is also a conditional distribution. We attempted to modify the structure of the model to control for this using a method inspired by Wooldridge's (2005) treatment of "initial conditions" in panel-data models, but this led to implausible results, so we stayed with our sample of entrants.

We estimate our model with enlisted samples from the WEX for each branch of service. The samples were drawn randomly and consist of approximately 3,000 entrants per year over the sample period. In most of our estimations, we use sample data for the 11-year period, 1990–2000. Even for the most recent cohort, year 2000, this allows us to observe first-term reenlistment and, for those leaving the active component, the decision to enter the reserves.

Basic Pay, RMC, and Retirement Benefits

We measure military pay for active-duty members by RMC. In estimating the model, we use RMC for FY 2004. RMC accounts on average for over 90 percent of the cash pay received by active-duty personnel. It is the sum of basic pay, basic allowance for subsistence (BAS), basic allowance for housing (BAH), and the federal tax saved because the allowances are not taxed. Although we use a common table of RMC, the amount of RMC received by an individual in a given period depends on the individual's pay grade. The pay grade depends on promotion probabilities from one grade to the next; these may vary from period to period and are service-specific. Thus, the actual distribution of RMC among service members at, say, YOS 8 of active duty, will differ across the services, depending on the differences in their promotion probabilities.

RMC is not suitable for reservists except when they are on active duty. Reservists who are drilling but not on active duty receive a subsistence allowance for their two drilling days per month and do not receive a housing allowance. Reservists on active-duty training receive rations and housing in kind only during the two weeks of training and receive either a partial housing allowance or a rate applied for married members, unless they are housed in contract housing off-base.

We measure a nonactivated reservist's military pay by basic pay for monthly drills (12 per year) and active-duty training (14 days per year). Drill pay is 1/30 of MBP for each drill period of four hours. A weekend typically has four drill periods, so drill pay for a weekend is 4/30 of MBP. Those participating in active training receive 14 days of basic pay plus housing and subsistence allowances. We include a partial housing allowance for single members living on base and a housing allowance for married members (known as BAH II) participating in active-duty training. Given years of service and grade, a reservist's base level of annual pay is approximately

$$(12 \times \text{weekend drill pay}) + \left(14 \times (\text{BAS} + \text{daily basic pay})\right)$$
$$+ (\%\text{married} \times \text{BAH II}) + (\%\text{single} \times \%\text{on base} \times \text{partial BAH}).$$

Some reservists receive special and incentive pays, but these are not included in our data.

The reserve retirement-benefit formula and the high-three active-duty retirement formula are programmed into our model; they are described in Appendix A. The present analysis assumes that reservists earn 80 points per year. The model focuses on behavior up to the age of 60, the age at which a reservist may begin receiving retirement benefits. We use life tables when determining the present value of active and reserve retirement benefits after age 60.

Promotions

We base promotion probabilities on tabulations of the WEX data. For each YOS and pay grade, we tabulate the percentage promoted to the next grade by the next period for each service. For instance, for Army

E-4s who have completed four years of service, we tabulate the percent promoted to E-5 by YOS 6. The promotion probabilities are in line with time-in-grade and time-in-service requirements for promotion (Williamson, 1999).

To implement promotions in the model, we assume that promotions occur at the beginning of the next period. The expected value of a military alternative in the next period, then, would be $V = (1 - p)V(g) + pV(g + 1)$, where p is the probability of promotion to the next grade.

Civilian Wages

We used a civilian median-wage profile with respect to total years of experience from Hosek et al. (2004). It is based on an analysis of March Current Population Surveys for 1983–2002, with samples limited to workers with at least 39 weeks of work in the previous year; most of these workers worked the entire year (52 weeks). The wages were deflated by the CPI urban deflator and put into 2004 dollars. About 90 percent of enlistees enter with a high-school education, and a majority of these obtain additional education while in service. Therefore, we used civilian earnings for workers with some college, i.e., more than high school but less than four years of college.

High-Year-of-Tenure Rules

Each branch of service has high-year-of-tenure rules limiting how long a service member may remain in a pay grade. The service may dismiss members who reach the high year of tenure in their grade. The rules have varied over time, relaxing somewhat when retention rates were low and tightening somewhat in the past few years, but on the whole, they have been fairly constant. The services do not dismiss all service members who reach a high year of tenure. We used WEX data to tabulate the percentage of service members who, despite having reached a high year of tenure, were still in service two years later and four years later and whether they had been promoted. We used these percentages when implementing the high-year-of-tenure rules in our model. Specifically, an individual who reached a high year of tenure faced a probability of being separated from the service and 1 minus the probability of being allowed to stay in the service. (See Appendix C for additional details.)

High Year of Tenure

Table C.1 shows the high year of tenure, by grade, used in our analysis of the active component. As shown in Table C.1, the high year of tenure for an E-4 was the 14th year, except in the Marine Corps, where it was the eighth year.

Table C.2 shows the percentage of service members in an active component who reached a high year of tenure as indicated in Table C.1 and the percentage of those their status one period (two years) later. The statuses are separated, same grade, higher grade, and lower grade. The percentage reaching a high year of tenure tends to increase with pay grade. The percentage of service members who separate from the military within two years after reaching a high year of tenure also tends to be higher at higher grades. However, depending on the service and grade, roughly 10–30 percent are still in the military, and some of them have been promoted. The findings in Table C.2 provide a basis

Table C.1
High Year of Tenure in Active Components

Grade	Army	Navy	Air Force	Marine Corps
E-4	14	14	14	8
E-5	20	20	20	14
E-6	24	22	22	20
E-7	28	26	26	26
E-8	30	28	28	28

Table C.2
Status Two Years After Reaching a High Year of Tenure (percent)

Service and Grade	Reached High Year of Tenure	Separated	Same Grade	Higher Grade	Lower Grade
Army					
E-4	0.47	57	14	28	0
E-5	2.56	89	9	2	0
E-6	3.36	89	7	5	0
E-7	3.42	71	19	9	—
E-8	11.25	100	—	—	—
Navy					
E-4	0.42	35	14	50	1
E-5	4.93	89	8	4	0
E-6	18.28	81	14	5	0
E-7	18.49	78	16	6	—
E-8	31.75	71	17	12	—
Air Force					
E-4	0.34	45	16	38	0
E-5	6.35	92	3	4	0
E-6	7.88	79	6	16	0
E-7	13.32	94	4	2	—
E-8	13.58	87	5	8	0
Marine Corps					
E-4	9.32	44	6	49	1
E-5	3.77	55	9	36	0
E-6	16.63	93	3	3	0
E-7	1.79	90	2	8	—
E-8	7.97	56	17	27	—

for our modeling assumption that, when a service member reaches a high year of tenure, there is some chance of continuing in the service. This chance may reflect our imperfect knowledge of the high-year-of-tenure policy in place each year, as well as the possibility that some service members at a high year of tenure have been selected for promotion but not yet promoted.

Comparisons with Different Discount-Rate Assumptions

Tables D.1, D.2, and D.3 replicate Table 5.2 in Chapter Five but assume different personal discount rates. Table 6.2 in Chapter Six compared retirement wealth under the QRMC alternative versus the current system from the standpoint of the individual member leaving at different years of service, assuming a personal discount rate of 15 percent—the rate that best fits our data with our model in terms of the active-duty profile. Tables D.1, D.2, and D.3 recompute the figures in Table 6.2 assuming a personal discount rate of 5 percent, 10 percent, and 12.5 percent, respectively. As discussed earlier, the key conclusions about the relative value of the current system versus the QRMC alternative to the individual change when alternative discount rates are used.

Table D.1
Present Value of Funds Available to Leaving Service Member, Current System, Option 1 Immediate, and Option 1 Deferred, Discount Rate = 5% (thousands of dollars)

	Option	E-5 at YOS 10	E-7 at YOS 20	E-8 at YOS 24	E-9 at YOS 30	O-3 at YOS 10	O-5 at YOS 20	O-6 at YOS 24	O-6 at YOS 30
Current system	Annuity payment	0	20	29	45	0	39	56	75
	PDV to 57	0	203	296	465	0	396	572	763
	PDV at 57	0	207	302	474	0	403	583	778
	PDV of PDV 57	0	87	155	331	0	169	299	543
	PDV of annuity	0	290	451	797	0	564	871	1,306
Option 1 Immediate	Annuity payment	0	3	10	30	0	6	20	48
	PDV to 57	0	35	99	178	0	68	190	292
	PDV at 57	0	40	134	393	0	77	260	664
	PDV of PDV 57	0	17	69	274	0	32	133	450
	PDV of annuity	0	52	172	470	0	102	332	771
	DC	11	40	57	93	22	76	108	169
	Separation pay	0	67	98	0	0	131	190	0
	Gate pay	0	10	12	15	0	23	27	34
	Total	11	169	339	579	22	332	657	973
	Total, if deferred	29	227	364	530	58	445	705	894

Table D.2
Present Value of Funds Available to Leaving Service Member, Current System, Option 1 Immediate, and Option 1 Deferred, Discount Rate = 10% (thousands of dollars)

	Option	E-5 at YOS 10	E-7 at YOS 20	E-8 at YOS 24	E-9 at YOS 30	O-3 at YOS 10	O-5 at YOS 20	O-6 at YOS 24	O-6 at YOS 30
Current system	Annuity payment	0	20	29	45	0	39	56	75
	PDV to 57	0	137	200	315	0	267	387	516
	PDV at 57	0	117	171	269	0	229	330	441
	PDV of PDV 57	0	20	43	129	0	38	84	211
	PDV of annuity	0	157	243	443	0	306	470	726
Option 1 Immediate	Annuity payment	0	3	10	30	0	6	20	48
	PDV to 57	0	25	76	154	0	48	146	253
	PDV at 57	0	26	87	255	0	50	169	418
	PDV of PDV 57	0	4	22	122	0	8	43	200
	PDV of annuity	0	29	100	286	0	57	192	470
	DC	11	40	57	93	22	76	108	169
	Separation pay	0	67	98	0	0	131	190	0
	Gate pay	0	9	11	13	0	21	24	30
	Total	11	145	266	393	22	285	514	669
	Total, if deferred	14	144	239	294	27	284	444	506

Table D.3
Present Value of Funds Available to Leaving Service Member, Current System, Option 1 Immediate, and Option 1 Deferred, Discount Rate = 12.5% (thousands of dollars)

	Option	E-5 at YOS 10	E-7 at YOS 20	E-8 at YOS 24	E-9 at YOS 30	O-3 at YOS 10	O-5 at YOS 20	O-6 at YOS 24	O-6 at YOS 30
Current system	Annuity payment	0	20	29	45	0	39	56	75
	PDV to 57	0	127	185	291	0	247	357	476
	PDV at 57	0	105	153	240	0	204	295	394
	PDV of PDV 57	0	14	33	105	0	28	64	173
	PDV of annuity	0	141	218	396	0	275	421	649
Option 1 Immediate	Annuity payment	0	3	10	30	0	6	20	48
	PDV to 57	0	23	71	149	0	45	138	245
	PDV at 57	0	24	81	235	0	46	156	386
	PDV of PDV 57	0	3	17	103	0	6	34	169
	PDV of annuity	0	27	90	262	0	52	175	429
	DC	11	40	57	93	22	76	108	169
	Separation pay	0	67	98	0	0	131	190	0
	Gate pay	0	9	11	13	0	21	24	29
	Total	11	142	256	368	22	285	496	627
	Total, if deferred	14	168	216	265	25	269	417	459

Bibliography

Asch, Beth J., and James Hosek, *Military Compensation: Trends and Policy Options*, Santa Monica, Calif.: RAND Corporation, DB-273-OSD, 1999. As of June 17, 2008:
http://www.rand.org/pubs/documented_briefings/DB273/

———, *Looking to the Future: What Does Transformation Mean for Military Manpower and Personnel Policy?* Santa Monica, Calif.: RAND Corporation, OP-108-OSD, 2004. As of June 17, 2008:
http://www.rand.org/pubs/occasional_papers/OP108/

Asch, Beth J., James Hosek, and David S. Loughran, *Reserve Retirement Reform: A Viewpoint on Recent Congressional Proposals*, Santa Monica, Calif.: RAND Corporation, TR-199-OSD, 2006. As of January 17, 2008:
http://www.rand.org/pubs/technical_reports/TR199/

Asch, Beth J., James Hosek, and Craig Martin, *A Look at Cash Compensation for Active-Duty Military Personnel*, Santa Monica, Calif.: RAND Corporation, MR-1492-OSD, 2002. As of June 17, 2008:
http://www.rand.org/pubs/monograph_reports/MR1492/

Asch, Beth J., Richard Johnson, and John T. Warner, *Reforming the Military Retirement System*, Santa Monica, Calif.: RAND Corporation, MR-748-OSD, 1998. As of June 17, 2008:
http://www.rand.org/pubs/monograph_reports/MR748/

Asch, Beth J., and John T. Warner, *A Policy Analysis of Alternative Military Retirement Systems*, Santa Monica, Calif.: RAND Corporation, MR-465-OSD, 1994a. As of June 17, 2008:
http://www.rand.org/pubs/monograph_reports/MR465/

———, *A Theory of Military Compensation and Personnel Policy*, Santa Monica, Calif.: RAND Corporation, MR-439-OSD, 1994b. As of June 17, 2008:
http://www.rand.org/pubs/monograph_reports/MR439/

Ben-Akiva, Moshe, and Steven Lerman, *Discrete Choice Analysis: Theory and Application to Travel Demand*, Cambridge, Mass.: MIT Press, 1985.

Berndt, Ernst, Bronwyn Hall, Robert Hall, and Jerry Hausman, "Estimation and Inference in Nonlinear Structural Models," *Annals of Economic and Social Measurement*, Vol. 3–4, 1974, pp. 653–665.

Berkovec, James, and Steven Stern, "Job Exit Behavior of Older Men," *Econometrica*, Vol. 59, No. 1, January 1991, pp. 189–210.

Bicksler, Barbara A., Curtis L. Gilroy, and John T. Warner, eds., *The All-Volunteer Force: Thirty Years of Service*, Washington, D.C.: Brassey's, 2004.

Christian, John, *An Overview of Past Proposals for Military Retirement Reform*, Santa Monica, Calif.: RAND Corporation, TR-376-OSD, 2006. As of June 17, 2008:
http://www.rand.org/pubs/technical_reports/TR376/

Cooper, Richard N., *The President's Commission on Military Compensation: A Review*, Santa Monica, Calif.: RAND Corporation, P-6260, 1978.

Defense Science Board Task Force on Human Resources Strategy, *Report of the Defense Science Board Task Force on Human Resources Strategy*, Washington, D.C.: Office of the Under Secretary of Defense for Acquisition, Technology, and Logistics, February 2000. As of August 7, 2008:
http://handle.dtic.mil/100.2/ADA374767

Daula, Thomas V., and Robert A. Moffitt, "Estimating a Dynamic Programming Model of Army Reenlistment Behavior," in Curtis L. Gilroy, David K. Horne, and D. Alton Smith, eds., *Military Compensation and Personnel Retention: Models and Evidence*, Alexandria, Va.: U.S. Army Research Institute for the Behavioral and Social Sciences, February 1991, pp. 181–201. As of June 17, 2008:
http://handle.dtic.mil/100.2/ADA363401

DoD—*see* U.S. Department of Defense.

Fieveson, Alan H., "Explanation of the Delta Method," *Stata*, December 1999, revised May 2005. As of June 17, 2008:
http://www.stata.com/support/faqs/stat/deltam.html

Gotz, Glenn A., "Comment on 'The Dynamics of Job Separation: The Case of Federal Employees,'" *Journal of Applied Econometrics*, Vol. 5, No. 3, 1990, pp. 263–268.

Gotz, Glenn A., and John McCall, *A Dynamic Retention Model for Air Force Officers: Theory and Estimates*, Santa Monica, Calif.: RAND Corporation, R-3028-AF, 1984. As of June 17, 2008:
http://www.rand.org/pubs/reports/R3028/

Hosek, James, Michael Mattock, C. Christine Fair, Jennifer Kavanagh, Jennifer Sharp, and Mark E. Totten, *Attracting the Best: How the Military Competes for Information Technology Personnel*, Santa Monica, Calif.: RAND Corporation, MG-108-OSD, 2004. As of June 17, 2008:
http://www.rand.org/pubs/monographs/MG108/

Keane, Michael, and Kenneth Wolpin, "The Solution and Estimation of Discrete Choice Dynamic Programming Models by Simulation and Interpolation: Monte Carlo Evidence," *Review of Economics and Statistics*, Vol. 76, No. 4, November 1994, pp. 648–572.

Mattock, Michael, and Jeremy Arkes, *The Dynamic Retention Model for Air Force Officers: New Estimates and Policy Simulations of the Aviator Continuation Pay Program*, Santa Monica, Calif.: RAND Corporation, TR-470-AF, 2007. As of June 17, 2008:
http://www.rand.org/pubs/technical_reports/TR470/

McFadden, Daniel L., "Econometric Models of Probabilistic Choice," in Charles F. Manski and Daniel L. McFadden, eds., *Structural Analysis of Discrete Data with Econometric Applications*, Cambridge, Mass.: MIT Press, 1981, pp. 198–272.

Public Law 93-406, Employee Retirement Income Security Act of 1974, September 2, 1974.

Public Law 106-65, National Defense Authorization Act for Fiscal Year 2000, October 5, 1999.

Rostker, Bernard, *Reforming the Military by Lengthening Military Careers*, unpublished RAND research, Santa Monica, Calif.: RAND Corporation, 2005.

Stern, Steven, "Simulation-Based Estimation," *Journal of Economic Literature*, Vol. 35, No. 4, December 1997, pp. 2006–2039.

Schirmer, Peter, Harry J. Thie, Margaret C. Harrell, and Michael S. Tseng, *Challenging Time in DOPMA: Flexible and Contemporary Military Officer Management*, Santa Monica, Calif.: RAND Corporation, MG-451-OSD, 2006. As of June 17, 2008:
http://www.rand.org/pubs/monographs/MG451/

Train, Kenneth E., *Discrete Choice Methods with Simulation*, Cambridge, UK: Cambridge University Press, 2003.

U.S. Census Bureau and U.S. Department of Labor Statistics, Current Population Survey, various years. As of June 17, 2008:
http://www.census.gov/cps/

U.S. Department of Defense, *Report of the Seventh Quadrennial Review of Military Compensation*, Washington, D.C., August 21, 1992. As of June 17, 2008:
http://handle.dtic.mil/100.2/ADA265369

———, Defense Advisory Committee on Military Compensation, *The Military Compensation System: Completing the Transition to an All-Volunteer Force*, April 2006. As of April 29, 2008:
http://www.defenselink.mil/prhome/docs/dacmc_finalreport7_11.pdf

————, *Commission on the National Guard and Reserves: Transforming the National Guard and Reserves into a 21st Century Operational Force*, Arlington, Va., January 2008a. As of April 29, 2008:
http://www.cngr.gov/Final%20Report/CNGR_final%20report%20with%20cover.pdf

————, *Report of the 10th Quadrennial Review of Military Compensation*, Vol. 1: *Cash Compensation*, Washington, D.C., February 2008b. As of August 7, 2008:
http://www.defenselink.mil/prhome/docs/Tenth_QRMC_Feb2008_Vol%20I.pdf

Warner, John T., *Thinking About Military Retirement*, Alexandria, Va.: Center for Naval Analyses, CRM D0013583.A1/Final, January 2006. As of June 17, 2008:
http://www.cna.org/documents/D0013583.A1.PDF

Warner, John T., and Saul Pleeter, "The Personal Discount Rate: Evidence from Military Downsizing Programs," *American Economic Review*, Vol. 91, No. 1, March 2001, pp. 33–53.

Williamson, Stephanie, *A Description of U.S. Enlisted Personnel Promotion Systems*, Santa Monica, Calif.: RAND Corporation, MR-1067-OSD, 1999. As of June 17, 2008:
http://www.rand.org/pubs/monograph_reports/MR1067/

Wooldridge, Jeffrey M., "Simple Solutions to the Initial Conditions Problem for Dynamic, Nonlinear Panel Data Models with Unobserved Heterogeneity," *Journal of Applied Econometrics*, Vol. 20, No. 1, January–February 2005, pp. 39–54.